THE UNIVERSE INSIDE YOU

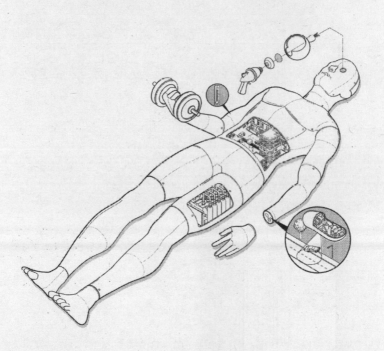

THE UNIVERSE INSIDE YOU

The Extreme Science of the Human Body from Quantum Theory to the Mysteries of the Brain

Brian Clegg

ICON

First published in the UK in 2012 by Icon Books Ltd

This edition published in the UK and USA in 2013 by
Icon Books Ltd, Omnibus Business Centre,
39–41 North Road, London N7 9DP
email: info@iconbooks.net
www.iconbooks.net

Sold in the UK, Europe and Asia
by Faber & Faber Ltd, Bloomsbury House,
74–77 Great Russell Street,
London WC1B 3DA or their agents

Distributed in South Africa
by Book Promotions,
Office B4, The District, 41 Sir Lowry Road,
Woodstock 7925

Distributed in Australia and New Zealand
by Allen & Unwin Pty Ltd,
PO Box 8500, 83 Alexander Street,
Crows Nest, NSW 2065

Distributed in Canada by
Penguin Books Canada,
90 Eglinton Avenue East, Suite 700,
Toronto, Ontario M4P 2YE

Distributed to the trade in the USA
by Consortium Book Sales and Distribution
The Keg House, 34 Thirteenth Avenue NE, Suite 101
Minneapolis, Minnesota 55413-1007

ISBN: 978-184831-504-4

Typeset in Melior by Marie Doherty

Printed and bound in the UK by
CPI Group (UK) Ltd, Croydon, CR0 4YY

Contents

CONTENTS

List of illustrations

Acknowledgements

For Gillian, Chelsea and Rebecca.

My grateful thanks to Simon Flynn, Duncan Heath, Andrew Furlow, Harry Scoble and all at Icon for their help and support.

I'd also like to thank all the real scientists who have answered my idiot questions, including Dr Henry Gee, Professor Stephen Curry, Professor Dan Simons, Professor Arnt Maasø, Dr Mike Dunlavy, Professor Günter Nimtz, Professor Friedrich Wilhelm Hehl and Dr Jennifer Rohn.

Introduction

We are used to science being something remote, performed by experts in laboratories full of strange equipment or using vast and highly technical machinery like the Large Hadron Collider. But we all have our own laboratories in the form of our bodies – hugely complex structures that depend for their functioning on all of the many facets of science and nature.

In this book you will use the workings of your body as a tool to explore the science of the universe. Some of that exploration will be very close to home, while for some of it you will necessarily journey away from your body, to the heart of stars and beyond. These tangents always have a point, illustrating the fundamental science that underlies reality, and we will always, in the end, return to that most miraculous of constructs that is the human body.

Brian Clegg, 2012

1. In the mirror

Stand in front of a mirror, preferably full length, and take a good look at yourself. Not the usual glance – really take in what you see. You may become a little coy at this point. It's easy to start looking for imperfections, noticing those extra centimetres on the waistline, perhaps. But that's not the point. I want you to really look at a human being.

In this book you are going to use the human body, your body, to explore the most extreme aspects of science. It's all there. Everything from the chemistry of indigestion to the Big Bang and the most intractable mysteries of the universe is reflected in that single, compact structure. Your body will be your laboratory and your observatory.

You can look at the whole body, treating it as a single remarkable object. A living creature. But you can also plunge into the detail, exploring the ways your body interacts with the world around it, or how it makes use of the energy in food to get you moving. Zoom in further and you will find somewhere between ten and 100 trillion cells. Each cell is a sophisticated package of life, yet taken alone a single cell is certainly not *you*. Go further still and you will find complex chemistry abounding – you have a copy of the largest known molecule in most of your body's cells: the DNA in chromosome 1.

Continue to look in even greater detail and eventually you will reach the atoms that make up all matter. Here traditional numbers become clumsy; a typical adult is made up of around 7,000,000,000,000,000,000,000,000,000 atoms. It's much easier to say 7×10^{27}, simply meaning 7 with 27 zeroes after it. That's more than a billion atoms for every second the universe is thought to have existed.

There's a whole lot going on inside that apparently simple form that you see standing in front of you in the mirror.

On reflection

In a moment we'll plunge in to explore the miniature universe that is you, but let's briefly stay on the outside, looking at your image in the mirror. Here's a chance to explore a mystery that puzzled people for centuries.

Stand in front of a mirror. Raise your right hand. Which hand does your reflection raise?

As you'd expect from experience, your reflection raises its left hand.

Here's the puzzle. The mirror swaps everything left and right – something we take for granted. Your left hand becomes your reflection's right hand. If you close your right eye, your reflection closes its left. If your hair is parted on the left, your reflection's hair is parted on the right. Yet the top of your head is reflected at the top of the mirror and your feet (if it's a full-length mirror) are down at the bottom. Why does the mirror switch around left and right, but leave top and bottom the same? Why does it treat the two directions differently?

Here's a chance to think scientifically. There are three things influencing how the mirror produces your image. The way light travels between you and the mirror, the way that you detect that light (with your eyes) and, finally, the way that your brain interprets the signals it receives.

We will explore all of these aspects of your body in more detail later in the book, but one significant point may leap out immediately as you think about the process of seeing your reflection. Your eyes are arranged horizontally. You have a left and a right eye, not top and bottom eyes. Could this be why the switch only happens left and right?

Sadly, no. It's a pretty good hypothesis, but in this case it's wrong. That's not a bad thing; much of our understanding of science comes from discovering why ideas are wrong. Let's try a little experiment that will help clarify what is really happening.

Experiment – On reflection

Hold up a book (or magazine) in front of you, closed, with the front cover towards you. Look at the book in the mirror. What do you see? Be as precise as possible. List everything that you can say about the reflected book. Does this help explain why the mirror works the way it does?

Do try this yourself first, but here's what I see:

- The book in the mirror is printed in mirror writing, swapped left to right.
- The reflected book is as far behind the mirror as my book is in front of it.
- The book's colours are the same in the mirror as they are on my side.
- The front cover of the book in the mirror is the back cover of my book.

Just take a look at that last statement. If I simply consider the book in the mirror to be an ordinary book then, as I look at it, my book's back cover has become the mirror book's front cover. Lurking here is the explanation of the mirror's mystery. It doesn't swap left and right at all. It swaps back and front.

In effect, what the mirror does is turn an image inside out. The back of my book becomes the front of the book in the mirror. Put the book down and look at your own reflection again. Imagine that your skin is made of rubber and is detachable. Take off that imaginary skin, move it straight through the mirror and, *without turning it round*, turn it inside out. The point of your nose, which was pointing into the mirror is now pointing out of the mirror. The parts of you that are nearest the mirror are also nearest in the reflection. Your entire image has been turned inside out.

In reality there is no swapping of left and right, so you don't have to explain why the mirror handles this differently from top and bottom. The reason we have the illusion of a left-right switch is down to your brain. When you see your reflection in a mirror your brain tries to turn the reflection into you. It makes a fairly close match if it rotates you through 180 degrees and moves you back into the mirror. This half turn flips left and right. But the key thing to realise is that it's not the mirror that performs a swap of left and right, it is your brain, trying to interpret the signals it receives from the mirror.

Now, with the mirror's mystery solved, let's start our exploration of the universe by taking a look at a single, rather unusual part of your body. We are going to investigate a human hair.

2. A single hair

Take a firm hold of one of the hairs on your head and pull it out. No one said science was going to be entirely painless. If you want to make this less stressful, get a hair from a hairbrush. If you are bald, get hold of someone else's hair – but ask first! Now, examine what you've got. It's a long, very narrow cylinder, flexible yet surprisingly strong considering how thin it is.

Take as close a look at the hair as you can. If you can lay your hands on a microscope, use that, but otherwise use a magnifying glass.

That strand of hair is going to start us off on everything from philosophy to physics. Dubious about just how philosophical hair can be? Consider this: you are alive and that hair is an integral part of you (or at least it was until you pulled it out). Yet the hairs on your body are dead – they are not made up of living cells. The same is true of fingernails and toenails. So you are alive, but part of what goes to make 'you' is dead.

Remember that next time a TV advert is encouraging you to 'nourish' your hair. You can't feed hair. You can't make it healthy. It's dead. Deceased. It has fallen off its metaphorical perch. Worried that your hair is lifeless? Well, don't be. That's how it is supposed to be. It's quite amazing just how many hair products are advertised using the inherently meaningless concept of 'nourishing'.

We're talking about a single hair, but of course you have (probably) got many more than one on your head. A typical human head houses around 100,000 hairs, though those with blonde hair will usually have above the average, and those with red hair rather fewer. Looking at that individual hair, the colour that provides this distinction doesn't stand out the same way it does on a full head of hair, but it's still there.

The colours of nature

The colour in hair comes from two variants of a pigment called melanin. One, pheomelanin, produces red colours. Blonde and brown hair colourings are due to the presence of more or less of the other variant of the pigment, eumelanin. This is the original form of hair pigment – red hair is the result of a mutation at some point in the history of human development.

As we become older, the amount of pigment in our hair decreases, eventually disappearing altogether. Grey and white hairs don't have any melanin-based pigment inside. In effect they are colourless, but the shape of the hair and its inner structure has an effect on the way that the light passes through it, producing grey and white tones.

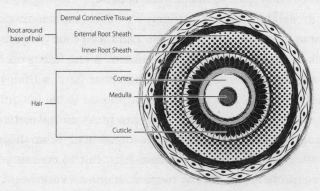

Cross-section of a human hair

The inner structure of hair isn't particularly obvious when you hold a single strand in your hand and look at it with the naked eye, but under a microscope it becomes clear that there is more going on than just a simple filament of uniform material. In effect your hairs have three layers: an inner one that is mostly empty, a middle one (the cortex) that has a complex structure that holds the pigments and can take in water to swell up, and an outer layer called the cuticle which looks scaly under considerable magnification, and which has a water-resistant skin.

On the end of the hair, where you have pulled it out of your scalp, there may be parts of the follicle, the section of the hair usually buried under your skin. The follicle is responsible for producing the rest of the structure and is the only part of the hair that is alive.

Dyeing to be attractive

The idea that the colouring of your hair is produced by melanins assumes it has its natural hue, but many of us have changed our hair colour using dyes at one time or another. Dyes use a surprisingly complex mechanism to carry out the superficially simple task of changing a colour. It's not like slapping on a coat of paint – the process of dyeing hair owes more to the chemist's lab than the beauty salon.

In a typical permanent dyeing process, a substance like ammonia is used to open up the hair shaft to gain access to the cortex. Then a bleach, which is essentially a mechanism for adding oxygen, is used to take out the natural colour. Any new colouration is then added to bond onto the exposed cortex. Temporary dyes never get past the cuticle; they sit on the outside of the hair and so are easily washed off.

Worrying about hair loss

Almost every human being has hairs, but compared with most mammals we are very scantily provided. Not strictly in number – we have roughly the same number of hairs as an equivalent-sized chimpanzee – but the vast majority of these hairs are so small as to be practically useless.

Next time you are cold or get a sudden sense of fear, take a look at the skin on your arms. You should be able to see goose bumps or goose pimples. This hair-related (indeed, hair-raising) phenomenon links to the fact that our ancestors once were covered in a thick coat of fur like most other mammals.

When you get goose bumps, tiny muscles around the base of each hair tense, pulling the hair more erect. If you had a decent covering of fur this would fluff up your coat, getting more air into it, and making it a better insulator. That's a good thing when you are cold, at least if you have fur – now that we've lost most of our body hair, it just makes your skin look strange without any warming benefits.

Similarly, we get the bristling feeling of our hair standing on end when we're scared. Once more it's a now-useless ancient reaction. Many mammals fluff up their fur when threatened to make themselves look bigger and so more dangerous. (Take a dog near to a cat to see the feline version of this effect in all its glory. The cat will also arch its back to try to look even bigger.) Apparently we used to perform a similar defensive fluffing-up, but once again the effect is now ruined by our relatively hairlessness. We still feel the sensation of having our hair stand on end, but get no benefit in added bulk.

Our lack of natural hairy protection struck me painfully when out walking my dog recently. It was a cold day and I was under-dressed for the weather in a short sleeved shirt. I was shivering and my trainers were soaked from the wet grass, so that I squelched as I walked. When passing through the fence from one field to the next, I managed to brush against a rampant clump of nettles, stinging both my arms.

But the dog, with her thick fur coat and hard padded feet, was impervious to both the weather and the vegetation. She seemed much better prepared to survive what nature could throw at her than I was.

I wondered why human beings are so badly equipped to cope with the discomforts and dangers of the natural world. We know that our distant ancestors had good, thick coats of protective fur, just as the apes still do today. (Present-day apes like chimpanzees and gorillas aren't our ancestors, but it's a mistake that's still often made in describing them.) It seems counter-intuitive that the early humans should have lost that helpful fur.

Of course, it's a misunderstanding to think that evolution has our best interests in mind. Evolution doesn't *have* a mind, or any concept of what is good or bad for us. Evolution usually works by gradual selection of subtle variants that enhance the survival and reproduction capabilities of individual members of species. It doesn't take an overview and think 'That's good, I'll keep that'.

Even so, it seemed unlikely that there was any evolutionary benefit in losing the warmth and protection of that natural fur coat.

Just because evolution deals us a set of cards it doesn't mean that everything we receive in our genetic hand is beneficial. There doesn't have to be an obvious evolutionary advantage just because we have developed a certain trait. It's just as likely to be a side effect of another evolutionary development. For example, many birds have wings that are easily snapped, because the bones are thin and hollow. Having weak bones isn't a good thing in itself – on the contrary, it's bad for survival. However, it is necessary to reduce the bird's weight enough for it to be able to fly.

There are various possibilities as to why it made evolutionary sense to lose the majority of our hair. It might have been due to the need to sweat more as our ancestors moved from the forest to the savannah – it's easier to sweat with less hair, exposing more skin for sweat to evaporate. Equally it could have been a response to the increase in parasites (though all the great apes are afflicted with these). Most exotically it has been suggested that early humans were partly aquatic, and less body hair made for a sleeker swimmer (though many semi-aquatic mammals are hairy). But the explanation that works best for me is that the loss was an accidental side effect, like those precariously thin bird bones.

To make allies, lose your hair

Around 100,000 years ago our distant ancestors went through the final changes that made them into modern humans. That was the end of our evolution to date. We are the same biological species now as they were back then. There have been plenty of tiny changes at the genetic level, but as a species we are essentially the same. We have the same potential for physical strength, for longevity, for attracting the opposite sex, for thinking and more.

Those many thousands of years ago, our predecessors had undergone huge evolutionary changes from the common ancestor they shared with chimpanzees and the other great apes. The pre-humans had lost most of their hair, leaving a delicate, thin skin exposed. They had shifted from a four-legged gait to walking upright. Their brains had grown out of all proportion with their bodies, leaving them bulgy-headed and top heavy (quite possibly unattractive features at the time). Their mouths had become smaller, making their teeth less effective as a biting weapon. The big toe had ceased to be an opposing digit that could be used to grip a tree branch.

Taken together, these alterations made the pre-humans more vulnerable to attack by predators. Their naked, unprotected skin was pathetically easy for claws and teeth to rip through. Compared with the smooth, four-footed pace of other apes, their tottering movements on two legs were painfully clumsy – a rabbit could easily outrun this strange unstable creature. The adaptations that came through in pre-humans don't seem to make any sense except as side effects. Put them alongside the change of behaviour that may have triggered them, and they were an acceptable price to pay.

These physical modifications of pre-humans are likely to have been an indirect result of an environmental upheaval. As the global climate underwent violent change, our ancestors were pushed out of the protective forests into the exposed world of the savannah. Facing up to starkly efficient predators, they were forced to change behaviour or become extinct. Back then, most pre-humans could not function well in large groups. This is still the case with most of our close relatives. The chimpanzee, for example, is incapable of forming large, cooperative bands. Get more than a handful of males together and the outcome is bloody carnage as battles for supremacy break out.

The pre-humans who first straggled onto the savannah around five million years ago were probably much the same. But the fast, killing-machine predators of the day – from the terrifying sabre-toothed dinofelis and the lion-sized machairodus to the more familiar hyena – made sure that things changed. The most likely pre-humans to survive were those with a natural tendency to cooperate. Our ancestors began to live in larger groups, giving them the ability to take on a predator and win, where a small band would be torn to pieces. And this change of behaviour may well have brought with it as side effects all the physical oddities that we observe in modern man.

The characteristics that repressed aggression and enhanced the ability to cooperate are typical of juvenile apes. Our primate cousins' inability to function in large groups only appears with maturity. The individuals amongst our predecessors who were more likely to survive on the savannah, those with the immature ability to get on with their fellows rather than tear them to pieces, were also the least physically developed. The eventual outcome was lack of hair on most of the body, a large head, a small mouth and even the upright stance – all features of the early part of the primate lifecycle that have normally disappeared by the time an individual matures.

As an aside, this mechanism of selecting for cooperative behaviour and getting an infant-like version of the animal is something humanity has since managed to produce repeatedly in its domestic animals. The dog, for example, has much more in common with a wolf cub than with the mature wolf that it was bred from. This is not just a matter of theory. In a fascinating long-term experiment between the 1950s and the 1990s, Russian geneticist Dmitri Belyaev selectively bred Russian silver foxes for docile behaviour and showed just how early man managed to turn the wolf into a dog.

Over 40 years – an immensely long experiment, but no time in evolutionary terms – the fox descendants began to resemble domesticated dogs. Their faces changed shape, becoming more rounded. Their ears no longer stood upright, but drooped down. Their tails became more floppy. Their coats ceased to be uniform in appearance, developing colour variations and patterns. They spent more time playing, and constantly looked for leadership from an adult. As they became more cooperative, they took on the physical appearance and the behaviour patterns of overgrown fox cubs.

To get back to humans, in the process of becoming more cooperative, and so more infantile (neotenous in the scientific jargon), the pre-humans lost the majority of their hair, leaving us with the largely hairless appearance we have today. Except, of course, on our heads.

Head hair can be lush in the extreme, and unlike the rest of our body hair (and that of other mammals) it just keeps on growing.

As with our general lack of hair, there are several possible explanations for this. It's quite possible that originally all our hair stayed at a roughly fixed length, but over time natural selection moved us towards head hair that continued to grow. This could be because those with a mutation causing head hair to keep growing had better protected brains. Or it could have been a side effect of wearing clothes, leaving the head most in need of furry protection. Or it could have provided a shield against the full impact of the noonday Sun, which can be formidable (as anyone with a bald patch can testify). Or there might be another, quite different explanation.

Tracing back the 'reason' for an evolutionary trait like this is notoriously difficult because we can't directly observe what happened or do an experiment to test a particular theory. It's a bit like news analysis saying that the stock market fell 'because of lack of confidence in the government', or for some other reason. No one really knows for certain why the market reacted this way, and similarly no one can prove why humans developed a particular trait. It is inevitably a matter of conjecture.

Lost in space

But given that we are now largely hairless, in some circumstances, clothing is a survival essential. Whether you are venturing under the sea or to the North Pole, your clothing is part of your equipment. And perhaps the greatest example of clothes-as-protection is when someone is out in space. Your body was never intended to be exposed to the extremes of space. The temperature is impossibly cold, as low as −270°C. There is no atmosphere. It's literally like nothing on Earth. Yet astronauts regularly make spacewalks protected only by specialist clothing.

It is possible to survive in space briefly without the right protection. Hollywood loves showing what would happen to a human being exposed unprotected, and can get it wonderfully wrong. The most ludicrous example is in the 1990 Arnold Schwarzenegger movie *Total Recall*, based on a Philip K. Dick story, where, expelled from the protected environment of a city on Mars, human beings inflate grossly before their heads explode messily.

Mars actually has a slight atmosphere (around one per cent of Earth's atmospheric pressure), and even in space this sort of inflation and explosion caused by low pressure isn't going to happen. There would be some discomfort as gas escaped from body cavities, but there is no danger that your head would inflate like a balloon.

It is true, though, that you would experience some liquids boiling. The lower the pressure, the lower the boiling point of anything, and in space – with no pressure to speak of – you will get an unpleasant drying up of the eyes as water boils away. Some fiction assumes your blood will boil in your veins, too – a horrible way to go – but according to NASA the pressure of your skin and circulatory system is enough to stop this happening.

Another worry is that you would freeze instantly in the very low temperatures of space. But bear in mind how a vacuum flask keeps its contents piping hot. Heat can only travel through a vacuum as light. We get our heat from the Sun in the form of light, which can happily cross empty space. Admittedly our bodies do glow with infrared – they do give off a degree of (invisible) light. But most of the heat we usually lose is passed on by conduction. The heat in our skin – atoms jiggling around with thermal energy – is passed on to the atmosphere, so our atoms jiggle a bit less, and the atmospheric atoms jiggle a bit more. That can't happen in a vacuum.

You would lose heat, but not very quickly. In practice, the thing that is going to kill you in space is simply the lack of air to breathe, and this will take a number of seconds. NASA has even experienced what would happen, when in 1965 a test subject's suit sprang a leak in a vacuum chamber. The victim (who survived) stayed conscious for around fourteen seconds in the airless chamber. According to NASA, the exact survival limit isn't known, but would probably be one to two minutes.

There's no doubt, then, that clothes can be important survival aids. Yet most of us, in everyday life, only have to cope with environments where plenty of other animals manage perfectly well with a bit of fur and some hardened skin on the feet. As naturists demonstrate, wearing clothes is often a social decision rather than an essential protection, and it's a decision we've been making for a long time. Woven cloth dates back at least 27,000 years – we know this because clay has been found at an ancient settlement at Pavlov in the Czech Republic with the imprint of woven cloth on its surface.

This isn't the oldest evidence for clothes we have, though. Bone needles have been found at Kostenki, a village in Russia, dating back around 40,000 years. These seem to have been used to stitch together animal skins to provide clothing. But the best clues to just how long we have been wearing clothes comes from the humble louse.

A lousy measurement

When Robert Hooke published *Micrographia* (see page 51), probably the most delight and revulsion came from his fold-out illustration of a louse. Seen magnified they are truly evil-looking parasites, specialist bloodsuckers that live on their host's skin, taking sips from the blood beneath. As many people with children at junior school know, the head louse is very fussy about sticking with its preferred environment around the base of head hairs. You don't find head lice straying to other parts of the body. But it does have a cousin that's less picky.

The human body louse evolved from the head louse between 50,000 and 100,000 years ago. We don't have ancient lice to work this out from, but this timing can be estimated by looking at the variations in the DNA of the two creatures – the more difference, the longer ago the division between head and body lice occurred.

This is of interest when thinking about the history of clothing because it's thought that the body louse was only able to develop once we started wearing clothes. Before then, the uncovered skin was too exposed. Interestingly, this 50,000 to 100,000 year timescale corresponds well with the timing of the move of humans out of Africa into colder climates, which could have been the spur that brought on the use of clothing.

Getting under your skin

Underneath your clothes, your body is covered in skin. Like hair, skin relies on melanin-based pigments to get its colouring. Also like hair, the outer layer of your skin is dead. The tiny flakes that contribute to the dust around your house fall off from this surface. Immediately below that dead layer called the stratum corneum (like the cornea in the eye, this 'corneum' comes from the Latin for horn, *cornu*) are two further layers, protective squamous cells and basal cells. The basal cells rise to the surface where they die, to form the outer coating, and they also play host to a different kind of cell, melanocytes, which produce skin pigments.

The more melanin the melanocytes pump out, the darker your skin. The normal state of your skin will have evolved to match the amount of ultraviolet in the light where your ancestors lived. Ultraviolet sits on the spectrum of light between visible light and x-rays – it is energetic enough to cause damage to the DNA inside your cells, if it can penetrate the outer layers of skin. Humans with a history of low exposure to ultraviolet – in the northern hemisphere – tend to lose melanin from the original African levels of their common ancestors.

This reduction in protection might not seem to have any advantage, merely adding risk if you get exposed to more sunlight (for example by emigrating to Australia), but in practice it was beneficial. This is because, despite the risk, the body needs some ultraviolet to get through, as it is used to produce the essential vitamin D. This is a vitamin that is relatively uncommon in food and that we need to avoid conditions like rickets. In northern

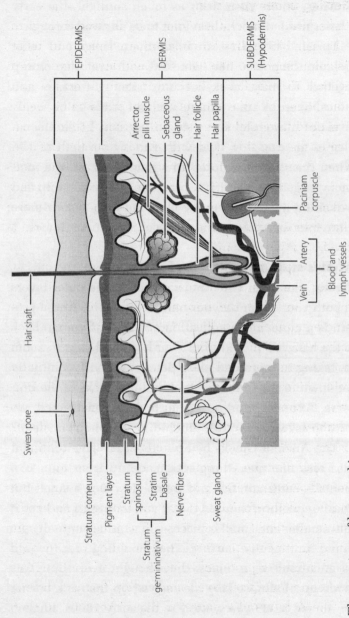

The structure of human skin

EPIDERMIS
DERMIS
SUBDERMIS (Hypodermis)

Arrector pili muscle
Sebaceous gland
Hair follicle
Hair papilla

Pacinian corpuscle

Artery
Vein
Blood and lymph vessels

Hair shaft

Sweat pore

Stratum corneum
Pigment layer
Stratum spinosum
Stratum basale
Stratum germinativum
Nerve fibre
Sweat gland

climates, where there isn't as much sunlight, the early settlers needed more ultraviolet to be allowed through.

This led to paler skin in northern areas, and what melanin the northerners were left with can often clump together to make dark patches, forming freckles and moles. Even in areas where sunlight tends to be weak, levels of ultraviolet can vary, so the skin has a mechanism – tanning – to deal with varying strength of UV. When the skin is exposed to strong sunlight, the melanocytes go into overdrive, producing more melanin and darkening the skin, thereby allowing it to absorb more ultraviolet and preventing damage to the lower layers.

What is stuff made of?

Keratin, the main structural material of the outer layers of both your skin and your hair, is a protein. And a protein is a molecule, a collection of atoms. If you go back to the hair you pulled from your head and start to zoom in, taking in more and more detail, you will eventually get down to the fundamental building blocks of the universe. To understand how your body is constructed, we have to ask what is 'stuff' (including your hair) made of?

The Ancient Greeks had two theories. The dominant idea was that everything was made up from four 'elements' – earth, air, fire and water. However, a small but vocal opposition thought that if you took stuff and cut it into smaller and smaller pieces you would eventually get to the limit of that cutting. The remaining piece would be uncuttable or *a-tomos*: they thought everything was made up of atoms. This idea stayed on the back burner for almost 2,000 years, until in the early 1800s, English scientist John Dalton devised modern atomic theory, suggesting that the different elements were made up of different types of small particle called atoms, each type unique to an element.

These elements were not the Ancient Greek four, but chemicals that could not be made out of others. Gases like hydrogen and oxygen, metals like iron and lead, and other substances like carbon and sulfur (for UK readers who think this word looks odd, this is now the standard worldwide chemical spelling for sulphur). Yet even at the start of the twentieth century, most scientists believed that atoms were just a useful concept to make chemistry work, rather than actual entities. It was only with work started by Albert Einstein in 1905 that atoms were finally considered to be real.

Battered by molecules

Atoms are a bit like small children – they are never entirely still. If you look at a glass of water sitting on a table, the water seems motionless. Yet within it, the water molecules are frantically (if randomly) rushing around. Einstein realised that an effect first observed by Scottish botanist Robert Brown in 1827 could be explained by the clumsiness of these energetic molecules.

Brown had spotted that the pollen grains of an evening primrose plant danced around in a drop of water when watched under a microscope. At first, Brown thought this was because there was some kind of life force in the pollen, but the same thing happened with ancient pollen and with stone dust and soot. It wasn't life in the pollen, but the activity of the water itself that created this 'Brownian motion'. Einstein realised that it was the water molecules randomly bashing into the pollen grains that caused the movement, and went on to give a mathematical basis for the theory. A little later, in 1912, French physicist Jean Perrin performed a wide range of experiments proving for the first time that atoms and molecules exist.

Remarkably, individual atoms can now both be manipulated and experienced visually. In 1989 a team working at IBM was the first to use a type of electron microscope that can manipulate as well as view, in order to move an individual atom. Two months later they arranged 35 atoms of the element xenon to spell out the initials IBM.

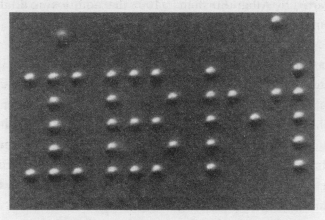

The letters IBM spelt out with xenon atoms
Photograph courtesy of Press Association Images

A little earlier, in 1980 Hans Dehmelt of the University of Washington isolated a single barium ion (an ion is just an atom with electrons missing, or extra electrons added, giving it an electrical charge). When illuminated by the right colour of laser light, that individual barium ion was visible to the naked eye as a pinprick of brilliance floating in space. You might argue that you couldn't 'see' the ion, just light that was reflected by it – but then that's all that ever happens when we see something.

Empty atoms and electromagnetic bottoms

The atoms that make up your body are not only very small, they are also mostly composed of empty space. If you could squeeze all the matter in your body together, removing the gaps, it would pack into a cube less than 1/500th of a centimetre on each side.

One of the wonders of the cosmos is the neutron star, a star in which the atoms have collapsed, losing all that empty space. In a single cubic centimetre of neutron star material – a chunk little more than the size of a sugar cube – there are around 100 million tons of matter. The entire star, heavier than our Sun, occupies a sphere that is roughly the size across of the island of Manhattan.

There is no danger of the atoms in you or your hair collapsing like a neutron star – without the massive gravitational pull of the star they remain stable. Collections of such atoms make up molecules like the keratin in your hair. The atoms stay together because of electromagnetism, one of the four forces of nature we will meet in more detail in Chapter 6. A molecule can be made up of a single element, like oxygen, the gas we breathe, which comes in molecules of paired atoms. Or it can be a compound, linking different elements, anything from simple sodium chloride – common salt – to the complex molecules, found in living organisms, like keratin.

The atoms that everything is composed from never touch each other. The closer together they get, the greater the repulsion between the electrical charges on their component parts. It's like trying to bring like poles of two intensely powerful magnets together. This is even the case when something *appears* to be in contact with something else. When you sit on a chair, you don't actually touch it. Your body floats an infinitesimal distance above, suspended by the repulsion between atoms.

It may be quite a while since you've played around with magnets. Get hold of a couple and remind yourself how remarkable the interaction between them really is.

Somehow the repulsion when you bring two of the same pole together seems more magical than attraction. Yet this is exactly what is happening every time one piece of matter 'comes into contact' with another. The interaction is electrical rather than magnetic, but it's a similar electromagnetic repulsion to the one you feel between the magnets that stops the atoms in your bottom slipping between the atoms of the chair.

Exploring an atom's innards

It wasn't long after atoms were proved to exist in 1912 that it turned out that the name was inaccurate. Atoms aren't 'uncuttable'. They have component parts. Scientists were already aware that there were negatively charged particles called electrons that could be pulled out of atoms. At first these were assumed to be scattered through a mass of positive material, like plums in a plum pudding (a description provided by British physicist J.J. Thomson). But a walrus-moustached New

Zealander working in Cambridge proved things were different.

Ernest Rutherford had the idea of firing other particles into an atom and seeing how they reacted – a bit like throwing a ball at an invisible structure and using the way the ball is influenced by what it hits to work out what that structure is like. The 'ball' he used was an alpha particle, a particle that had recently been discovered shooting out of radioactive elements. (It was later identified as the nucleus of a helium atom.) Alpha particles made tiny flashes when they hit screens painted with fluorescent material. By crouching in the dark it was possible for Rutherford's assistants to spot the paths of particles that were deflected to the sides as they were shot at a piece of gold foil.

With the kind of inspiration that makes all the difference in science, Rutherford and his team also looked for alpha particles that reflected off the atoms in the gold straight back towards the source – and occasionally one did. This was totally unexpected. Rutherford said it was like firing an artillery shell at a piece of tissue paper and having it bounce back at you. He realised it meant that atoms must have a small, very dense, positively charged core to repel the positive alpha particles. Rutherford established for the first time the familiar picture of an atom being like a solar system with a positive nucleus at the centre (he borrowed the word 'nucleus' from biology). The nucleus was the equivalent of the Sun and the negatively charged electrons were the planets of this tiny solar system.

Thomson's plum pudding was no more. The nucleus was so much smaller than the whole atom it was described as being like a fly in a cathedral, around 100,000 times smaller than the atom as a whole. The nucleus was made up of positively charged particles called protons, making up 99.9 per cent of the mass of the atom. For each proton an electron flew around the outside, balancing up the electrical charge, leaving the atom neutral.

But even this newly detailed picture wasn't quite good enough. In 1932 another particle was found in the nucleus – the neutron. This had a similar mass to the proton but no charge, and it helped explain a mystery. There exist different versions of the same element, called isotopes. They act in the same way chemically, but the atoms have different weights. The neutron explained this picture. The number of charged particles decide what element you have and how it reacts chemically. But different atoms of the same element can have varying numbers of neutrons in the nucleus, producing a range of weights.

No miniature solar system

When we imagine the atoms making up our bodies, this is the picture of what an atom 'really' is that many of us still have, but science has moved on since 1932. We now know that electrons don't fly around the nucleus like planets around the Sun – the solar system model just doesn't work. If it *was* an accurate picture, we'd have problems. When a charged particle is accelerated it gives off energy in the form of light. And orbiting is a form of acceleration. This is because acceleration doesn't mean a change of speed, which is the sense in which we tend to use the word, but rather a change of velocity.

Speed is just a number, like, say, 30 miles per hour. But velocity is more. It is speed *and* direction. So it might be 30 miles an hour, due north. Anything moving accelerates if any part of its velocity changes. So even if it is still going 30 miles per hour, it accelerates if it changes from heading north to heading east. If you think about an electron whizzing around in an atom like a miniature planet, it would always be changing direction, always accelerating. And that means it would lose energy as a burst of light and would plunge into the nucleus in a tiny fraction of a second. Every atom in the universe would instantly self-destruct.

Taking a quantum leap

The reason everything doesn't disappear in a flash is explained by quantum theory, the science of the very small. This tells us that the familiar picture of electrons as little particles, whizzing around in an orbit, is wrong. At any point in time, an electron isn't in a single position. Instead it is in many places around the atom simultaneously, each with different probabilities, only settling to a single location if it is observed. It's better to think of them as fuzzy clouds of probability around the outside of the atom. Of course, it's harder to draw a picture of that, so the old solar system model still features in many textbooks.

The electrons that produce this 'fuzz' on the outside of atoms can only exist with specific levels of energy. It's as if they run on rails. You can give them a boost of energy, in which case they will jump up to the next rail. But you can't give them an intermediate amount of energy; they can never end up positioned between rails. These fixed 'packets' of energy are called quanta, which is where the name 'quantum theory' comes from.

This also means that the term 'quantum leap' is used very strangely in everyday language. A quantum leap is the jump between one rail and the next one. It's the smallest possible change in the energy of an electron that there can be. So it is rather bizarre that in general usage it has come to mean a really significant transformation.

Usually, the energy to push an electron to a higher level (the 'rail' analogy is mine, it's not in general usage) is provided by light. Light carries energy (it's just as well that it does, because that's how the Sun's energy reaches us across the vacuum of space) and gives electrons those necessary boosts. Similarly when an electron drops down a level, it gives off light. But because the electron can only move from rail to rail, this energy is in packets – quanta. The light comes in packets – particles – which are called photons.

The charm of quarks

Your body is made of molecules, each containing atoms, each of which has an internal structure of protons, neutrons and electrons. But we know now that the old picture of protons and neutrons being the fundamental objects at the heart of an atom is also wrong. Protons and neutrons are both made of *truly* fundamental particles called quarks. There are quite a few types of quark, described by their 'flavour' (no, really). The different flavours include charm, strangeness, top and bottom, but the ones we're interested in are up and down. A proton is made of two up quarks and one down, while a neutron is two down quarks and one up.

This all works out in terms of electrical charge, because up quarks have a $\frac{2}{3}$ charge and down quarks have $-\frac{1}{3}$, resulting in a positive charge of 1 for the proton and no overall charge for the neutron. It sounds wrong that a particle should have a fraction of a charge, and quarks aren't really $\frac{1}{3}$ or $\frac{2}{3}$ of anything – they are the true units of charge. However, because protons and electrons are all that were known when the numbers were first established, we're stuck with thirds.

This odd name, quark, is often pronounced to rhyme with lark, but when American physicist Murray Gell-Mann dreamed up the idea, he wanted it to rhyme with cork. He came up with that 'kwork' sound without thinking about how to spell it. But then he came across a line in James Joyce's *Finnegans Wake*: 'Three quarks for Muster Mark!' The way quarks come in threes made the text very apt, so Gell-Mann adopted the spelling, even though it didn't fit his pronunciation.

The messy standard model

With quarks you have really reached the uncuttable – part of a bigger picture scientists use to describe all the particles that make up your body and the rest of the universe.

Physicists have produced something called the 'standard model', which describes everything we know in existence being based on around nineteen different fundamental particles. Twelve of these are matter particles, like quarks and electrons, plus some more obscure

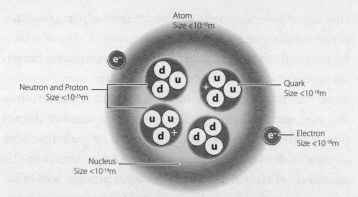

Atom
Size <10⁻¹⁰m

Neutron and Proton
Size <10⁻¹⁵m

Quark
Size <10⁻¹⁸m

Electron
Size <10⁻¹⁸m

Nucleus
Size <10⁻¹⁴m

The structure of an atom: to the scale shown for the nucleus,
the whole atom would be around ten kilometres across

variants found in nuclear reactions and collider experiments. Another five are special particles that carry forces. So, for instance, there's the photon which is both a particle of light and carries electromagnetic force from place to place.

There are also a couple of particles that may or may not exist – the graviton, which would be the particle that carried gravity, if gravity is indeed a force that comes in quantum chunks like the others (as yet this isn't fully supported by theory). And then there's the Higgs boson, the main target of the massive Large Hadron Collider at CERN, which is an elusive particle that is thought to give some of the other particles their mass.

To make things even more complex, each particle has an anti-particle. Antimatter sounds like something out of *Star Trek* (and in fact it is how the *Enterprise*'s engines are supposed to work), but it's very real. Antimatter is just like ordinary matter, but some of its properties, like the charge, are reversed. All twelve matter particles have an antimatter equivalent. So, for instance, the electron has the anti-electron, better known as a positron, which has a positive charge instead of a negative one.

If matter and antimatter are brought together they destroy each other and their mass is converted into energy. Because the energy in matter is quantified by Einstein's famous equation $E=mc^2$, and c, which is the speed of light, is a very big number, there's a whole lot of energy going on when matter and antimatter combine. A kilogram of antimatter, annihilating with an equivalent amount of matter, generates the equivalent of a typical power station running for around twelve years. (Depending on the antimatter used, there may be secondary particles called neutrinos produced in the reaction, which can reduce the energy output by half, but this is a relatively small consideration.) Antimatter is the most compact way to store energy that we have. It packs in 1,000 times more energy than nuclear fuel.

Although this zoo of different particles works pretty well at explaining everything that goes into the matter than makes up your hair – and everything else with mass or energy – it is a messy way of looking at things, and scientists would love to have a simpler picture to deal with the fundamentals of reality. For years physicists have been developing competing theories to achieve this, but as yet none is satisfactory.

Is it solid, liquid or gas?
Away from such theoretical considerations, an interesting question to ask when looking at your hair is what

kind of material it's made of. You were probably taught at school that all matter is solid, liquid or gas. As a hair clearly isn't liquid or gas it must be a solid, but something so flexible and pliant doesn't really fit with our immediate concept of a solid. We tend to think of a solid as rigid, not pliable. Sand is another good example of a substance that doesn't fit comfortably with simplistic classifications. Think of a fistful of sand – indubitably sand is made of solid particles, yet it runs through your fingers like a liquid.

We can get a better feel for these 'states of matter' from one of the few substances that we experience as solid, liquid and gas – water. From it we learn that the distinction between the three states of matter is twofold. The atoms are typically further away from each other and they are typically moving faster as we go from solid to liquid to gas. All atoms and molecules move, but in a solid they jiggle about in a well-established framework of bonds – electromagnetic links between molecules. In a liquid, there are still bonds, but they are less substantial and have no stable structure. In a gas the molecules act pretty well independently.

This makes it sound as if there is a continuum between states, but they are clearly defined. It's true that as a liquid, for example, molecules of water will constantly be escaping into gaseous form (evaporating), but if you want to turn a body of water into gas you have to heat it to the right temperature, the boiling point, and then give it extra heat (the 'latent heat of boiling') to remove the final bonds and let those molecules free.

The fourth state of matter

The science you were taught at school probably stopped with the Victorian idea of there being three states of matter, but in fact there are five states altogether. The fourth is one that you have experienced many times – it is a much more obvious state than gas – but because our school science is so strongly locked into the nineteenth-century worldview, even many adults don't know it exists, except as a label in relation to large screen TVs. It's plasma.

One potential point of confusion needs clearing up here, especially as our starting point in this book is your body. This plasma we are discussing has nothing to do with blood plasma. Blood plasma is the colourless liquid in which blood cells float. Plasma in the physics sense is the fourth state of matter, the one that comes beyond a gas. (Actually neither of the uses of the word are particularly good, as 'plasma' originally meant something formed or moulded, and both types of plasma are formless.)

It shows how badly plasma is understood that my dictionary defines it as being 'a gas in which there are ions rather than atoms or molecules'. Let's not worry about those ions for a moment, but note how fuzzy the dictionary writer's thinking was. To define a plasma like this is similar to calling a liquid 'a very dense gas with fluid properties'. A plasma is certainly more like a gas than a liquid, just like a gas is more like a liquid than a solid, but it is still something else; a different state of matter.

I said that plasmas are more obvious than gases because they are usually highly visible. The Sun is a huge ball of plasma. Every flame contains some plasma, although the flames we usually encounter are fairly cool in plasma terms, so usually consist of a mix of plasma and gas. Just as a gas is what happens to a liquid if you continue to heat it past a certain point, so a plasma is what happens to a gas if you continue to heat it far enough.

As the gas gets hotter and hotter, the electrons around the atoms in the gas get more and more energy. Eventually some have enough energy to fly off and leave the atom behind. Most atoms have a natural tendency either to lose or gain electrons. Atoms that easily lose electrons do so, and end up as a positively charged ion. Atoms that easily gain electrons hoover up the spare ones from the positive ions and end up as negatively charged ions. Ions are just charged atoms with either electrons missing or electrons added. A substance that has been heated so far that its atoms become ions is a plasma.

Plasmas are very common once you consider the universe as a whole. After all, stars are pretty big objects. It has been suggested that up to 99 per cent of the universe's detectable matter is plasma. In part this is because plasmas glow, so they are easier to spot. Although plasmas are gas-like, in not being hugely dense, they are very different from gases. For instance, gases are pretty good insulators, while plasmas are superb conductors.

Experiment – The state of custard

We usually think of materials changing state as a result of variations in temperature. Cool down water and it becomes ice. Heat up a piece of metal and it

---

becomes molten (liquid) metal. But pressure can also have a dramatic effect on some materials. Thixotropic non-drip paints change between gel form (a gel is a malleable solid) and liquid when stirred. But the most dramatic and fun demonstration of the effect of pressure on the state of matter is provided by custard.

Mix custard powder with water so you get a thick yellow liquid. Pour some into a bowl. Now put your finger and thumb into the liquid a few centimetres apart and squeeze them together. The liquid becomes a dry powder under the pressure of your fingers. As long as you keep the pressure up, it will stay solid – you can easily lift it out of the bowl – but as soon as you relax the pressure it will return to liquid and drip from your fingers.

This quality makes it possible to walk across the surface of a pool of custard. To see this in action, visit **www.universeinsideyou.com**, select *Experiments* and click on *Walking on Custard*.

Enter the condensate

The fifth state of matter is not custard, but it is just as strange. On a good day, scientists can come up with impressively snappy terms. 'Plasma' is pretty good. So are 'photon' and 'quark'. But all too often they come up with a name that no one in their right mind wants to say – try saying this one five times over very quickly. The fifth state of matter is a Bose–Einstein condensate.

This is a state down the other end of the temperature scale from a plasma. In fact, before we visit the condensate, it's worth just briefly thinking about temperature. What is temperature? It's how hot something is – fair enough. To heat things up we have to put energy into them. But what is happening as we do? The atoms or molecules in the material speed up. Even in a solid, atoms jiggle with energy. In a liquid they move about, while in a gas they positively rocket around the place.

When you use a thermometer to measure your body temperature (around 37°C), you are taking an average measure of the energy of movement in the particles that make you up. If you aren't sure about there being a difference in energy just because something's moving faster, imagine being hit by a tennis ball at 5 kilometres per hour, then at 500 kilometres per hour. The second one would hurt a lot more thanks to all that extra energy.

Unless you knew that temperature was about the movement of the atoms in a material, you might imagine that you could just cool things down indefinitely, getting colder and colder, as far as you liked, assuming your refrigeration mechanism was good enough. In practice, though, you can only slow down the atoms or molecules so much. Eventually they would stop. That temperature, unreachable in practice because quantum particles can never entirely stop, is absolute zero.

This ultimate low temperature is around −273.16°C. Scientists, however, often use a temperature scale that has the same size units as Celsius, but which starts sensibly with zero at absolute zero. This is the Kelvin scale, so 0°C is about 273 K on that scale. (For those who like pedantic detail, the units of the Kelvin scale are kelvins, with a small k, but the symbol is a capital K. Unlike

Farenheit and Celsius there are no 'degrees' – so the freezing point of water is 273.16 K, not 273.16°K.)

When materials get close to absolute zero, they begin to behave very strangely. Some substances become condensates (technically there are two variants, Bose–Einstein and Fermionic, but let's not worry about too much detail here). A condensate is a state of matter where the particles that make it up lose their individuality. This results in strange behaviours like superfluidity, where the substance has absolutely no resistance to movement. Superfluids climb out of containers of their own accord, because there is no resistance to the random movement of the molecules. If you start a superfluid rotating in a ring it will go on forever. Then there are superconductors, which have no electrical resistance.

The pièce de résistance of the condensate world is the way a Bose–Einstein condensate deals with light. Because the condensate is halfway between normal matter and light itself, it can interact with light in a strange way, slowing it to a crawl or even bringing it to an effective standstill. This weird mix of light and matter is called a 'dark state', a romantic name that well fits such an odd phenomenon.

Every kind of stuff

So that's five states of matter. Up at the top, plasma, a collection of high-energy ions. Next a gas, then a liquid, then a solid. Finally, at the extreme limits of cold, the Bose–Einstein condensate. It's easy to think of materials – stuff – as being rather ordinary and boring science. Yet there's a remarkable amount going on in that individual hair.

Look close enough and you have molecules, made of atoms. As we have seen, each atom has its nucleus of protons and neutrons (apart from hydrogen, which is so small that its nucleus is just a single proton) and its surrounding cloud of electrons. And each of the particles in the nucleus is made up of a triplet of quarks. These simple building blocks are responsible not only for the relatively straightforward structure of your hair, but for all the complexity that goes into your body.

You are what you eat

But where did the components of your body come from? Where were those atoms before they were incorporated into you? In previous centuries they were drifting around the planet, getting involved in all manner of reactions. There's an awful lot of carbon in your body, for instance. Where did it come from? Plants and animals, which in turn got theirs from other plants and animals. And if you go along the chain far enough you'll hit a vegetarian. So ultimately all that carbon came from plants. But where did they get it?

The air.

Plants have the wonderful ability to build themselves largely from air. We're used to carbon dioxide being treated as a bad guy because of its role as a greenhouse gas, but bear in mind that most of the carbon that gets incorporated into plants comes from the carbon dioxide they take out of the atmosphere. That's just as well, as they then pump out the waste oxygen, and that's the only reason we can breathe.

So prior to being in other animals and plants, some of your atoms were in the air. Some came from the ground and from water. Go back far enough, and many of them will have spent time in other people in history. There are so many atoms in a person (7×10^{27}) that after a while, many of them will be recycled in other human beings. Your body contains atoms from kings and queens, noble warriors and court jesters.

This is subtly different from the suggestion that every breath you take contains an atom or two that was breathed by Marilyn Monroe. The atmosphere moves around with sufficient vigour to mix those breaths into the whole and get the odd atom into your next intake of fresh air. But the atoms that made up Marilyn haven't had time to spread around the world and get into everyone's body. Some people will have them, but not everyone. Go a few hundred years forward in time, though, and it will be pretty certain that molecules of Marilyn will be in every person's body.

Components that pre-date the Earth

The atoms inside you have been circulating around on Earth since life began, well over three billion years ago. Fossils can be used to trace life back in rocks that were formed around 3.2 billion years ago, while the date can be pushed back a few hundred million years more on the basis of chemicals that suggest the existence of life. But before then, the atoms were still there. They didn't appear out of nowhere. The atoms that make you up were present when the Earth was formed 4.5 billion years ago (apart from a few that arrived since on meteors from outer space).

Before that they floated for aeons through space. Some have been around since the beginning of the universe. According to the Big Bang theory, our best idea of how the universe began, all of the hydrogen in the universe and some of the helium and lithium was created when the remnants of the Big Bang that formed the universe cooled down enough to stop being pure energy and formed matter. So the hydrogen in the water and organic molecules in your body date back to the very beginning of the universe.

After a while, some of this hydrogen clumped together, pulled by gravity, and formed stars, which burn in their youth by converting hydrogen, the lightest element, into the next element, helium. When most of the hydrogen is used up, helium too can be consumed, working up the elements all the way to iron. And this is where elements like the carbon and oxygen that are so important for life were forged.

Later still, some of those stars would become unstable and detonate in catastrophic explosions called supernovas. Ordinary stars don't have enough energy to make the elements that are heavier than iron, but supernovas have so much oomph that they can create elements all the way up to uranium, the heaviest of the naturally occurring elements.

This means that, quite literally, you are stardust. The atoms within the hair you hold, and within every part of your body, either came from the Big Bang – so are 13.7 billion years old – or from a star, which would make them between seven and twelve billion years old. The components of your hair – and every other part of you – are truly ancient. We tend to think of the universe explored by astronomers as very distant and not really connected with life on Earth. Yet every atom inside you was once out there, once part of the wider cosmos.

A sprinkling of stardust

This makes you rather special. Atoms are a rarity in the universe. There really aren't many of them out there. This might seem unlikely, considering all the stuff we see around us, let alone all the stars and galaxies in the universe, but it's a big place. It has been estimated that there are around 10^{80} atoms in the observable universe, that is, all of the universe it's possible to see. Distances in space tend to be measured in light years, the distance light covers in a year. As it travels around 300,000 kilometres per second, that makes a light year around 9.5 trillion kilometres. And the visible universe is about 90 billion light years across.

We say 'the visible universe' because no one is sure how big the universe is. However, there's quite a lot of evidence that suggests the universe came into existence about 13.7 billion years ago. So we can only see light that has been travelling for 13.7 billion years (slightly less, actually, but let's not worry about that). If everything stayed the same, that would make the visible universe about 27 billion light years across – but the universe has been expanding since it began. So the point the light set off from 13.7 billion years ago is now around 45 billion light years distant.

The universe is so big that if you distributed all the atoms in it evenly throughout space, there would only be one oxygen atom in about every 6,250 cubic metres. Just think of that in terms of your body. By far the biggest component of your body by mass is water. And most of water's mass is oxygen — so the biggest atomic component of your body is oxygen, which accounts for about 65 per cent of your mass. So, if all the matter in the universe was nice and evenly spread out, to provide the oxygen in you would require the contents of over 9×10^{30} cubic metres of space. A cube twenty million kilometres on each side; that's more than 50 times the distance to the Moon.

Think about that hair from your head once more. People take great pride in working out their genealogy over a few generations. If a country house has been owned by the same family for 400 years they consider themselves something special. But that hair you are holding has contents harvested from across space, with some of its atoms going all the way back to the Big Bang, and all of them well over five billion years old. That's what I call having ancestry.

Your hair, as you discovered earlier, is dead. But now it's time to move over to signs of life. And what's more suggestive of life than blood?

3. Locked up in a cell

You have no doubt cut yourself at some point and seen deep red blood well up from the wound. If you have a sterilised needle handy and would like to prick the ball of your thumb to take a closer look at a drop, feel free (provided you have no medical problems that make this dangerous) – but it's not essential. If you do decide to give this a go, as you stick the needle in, you may feel the urge to swear. And this isn't necessarily a bad thing.

Cursing the pain away

Research carried out in 2009 suggested that there is a good reason that we tend to yell expletives when we hurt ourselves. By comparing the effects of swearing against using everyday words, it was discovered that yelling swear words increased the ability to tolerate pain and decreased the amount of pain that was felt. According to the research, this relief didn't apply to men with a tendency to 'catastrophise'. As this word isn't in the *Oxford English Dictionary*, I'm not entirely sure what the scientists mean by it – I can only assume it's a tendency to be a drama queen.

The suggestion from the research was that swearing could break the link between fear of pain and the feeling of pain, reducing self-induced suffering. Whether this helps or not, there is a small amount of suffering required if you want to take a look at that drop of your blood.

A living liquid

Here is something very different from that lifeless hair. There is no doubt that your blood is active in a way your hair isn't. Yet it isn't easy to say at what point we move from something that is dead to something that's alive. Down at the level of atoms, the blood is no different from your hair, or, for that matter, from a rock. The specific mix of atoms may be different – there's a significant amount of iron in the blood, for example – but both are still made up of assemblies of atoms in the form of various molecules. Yet somehow the 'living' blood and the dead hair *are* different.

Deciding for certain whether or not something is alive is a surprisingly non-trivial task. Before you read on, see if you can list at least six things that distinguish something that is living from something that isn't.

At one time it was thought that there was a 'life force', a form of energy that was present in living things, but that wasn't there when they were dead. But this energy has never been detected and the life force is no longer taken seriously outside of pseudoscience and metaphor ('she looks full of energy today').

The signs of life

Instead, biologists look for seven signs that life is underway, known as life processes. Life is, in effect, defined by what it *does* rather than what it *is*. These seven processes are:

- Moving – even plants move over time; watch a sunflower follow the Sun

- Nutrition – consuming something to generate energy, whether that something is plants, animals or sunlight
- Respiration – the process by which energy is produced from the 'food' source, often but not always involving oxygen
- Excretion – getting rid of waste matter
- Reproduction – making new copies of themselves (often with variation) to continue the species
- Sensing – having some interaction with what is around, usually by detecting forms of energy
- Growth – though not a constant throughout life, all living things grow at some point in their development.

At the level of an organism – a plant or animal – the simple rule is that unless all these processes take place, whatever you are looking at isn't alive. Get all seven and you probably have a winner. Even here, though, making the 'dead or alive' call is not always totally clear. Take the virus that gave you an irritating sniffle some while back. You could regard a virus as a single-celled living thing. There are plenty of things that only have a single cell that definitely are alive, bacteria for example. Yet viruses fail on the reproduction test.

It's not that viruses don't reproduce – it is their reproduction that causes problems in your body. But the way they do it is to commandeer the mechanisms of their host's cells. In a sense, when you get a virus, it's *your* life that reproduces a virus, not its own. Many – though not all – biologists do not consider viruses to be alive, and in part it's the lack of life that makes it particularly difficult to get rid of them. You certainly won't get anywhere with an antibiotic, which is why taking these for colds and flu is a waste of time.

Are your cells alive?

It's harder still to be sure if something is alive when looking at a *part* of a living thing. With the exception of single-celled creatures, an organ or a cell taken in isolation certainly won't fulfil all the criteria – your heart can't reproduce, for example.

Life, as the biologists use the word, is a holistic term that really only works at the level of an organism. When an animal goes from being alive to being dead, we can't see immediate changes in every cell, though eventually they will come. So with this interpretation we can't say that a drop of your blood or a cell from your finger is alive. And yet there is so much more going on in there than was the case with your hair. We can't say that flesh and blood is dead in the same way as hair – parts of a living thing, like blood or a cell from the flesh of your thumb, will typically exhibit some (but not all) of the life processes.

I asked a cell biologist if she thought cells are alive, and she was very certain that they are. As she pointed out, 'This is never more evident than when you've had a bad day in the lab and you end up killing your cell cultures by mistake. Cells that are alive metabolise, and divide, and move around – if you film them with time-lapse microscopy, they are amazingly dynamic, quivering and pulsating and sending out probing little fingers (filopodia) and feet (lamellipodia); some cells even crawl around. And of course, they reproduce themselves, some endlessly, like immortal cancer cell lines. When cells die, they retract all their fingers and feet, and round up – their nucleus disintegrates and they sort of explode. Then they are utterly motionless, never to rise again. So in my view, this is clearly the difference between life and death!'

Blood cells are tricky in this respect. Unlike most of your cells they don't have a nucleus (more on this soon) and they just go with the flow with the circulation of your blood. However, they still play a hugely important and active role in keeping the rest of you alive.

A voyage through your bloodstream

If you look at a drop of blood that oozes from a pinprick on the end of your finger, it seems to be a dark red liquid with no particular bits and pieces in it – but get a smear of it on a microscope slide and it is packed with small objects. Some, the red cells, are like little lozenges, resembling tiny dried apricots. Their role is to carry oxygen from the lungs to the body tissues.

These cells are red because their main constituent is a protein called haemoglobin (your many different proteins are amongst the most important worker molecules in your body). Take away the water from red blood cells and 95 per cent of what's left is haemoglobin. This large molecule is excellent at binding onto oxygen to carry it around the body. Haemoglobin contains iron, and it is often thought that this causes the red colour just as it produces the red tint of rust, but the colouring is a coincidence. The iron atoms are bound in a ring of atoms called porphyrin, and it is this organic structure that provides the colouration. The red blood cells are produced in your bone marrow and typically whizz around your body every twenty seconds for around four months, along with trillions of others, before they are replaced.

The other familiar occupants of that drop of blood are the white blood cells. There are many types of these, acting as defence mechanisms and clean-up operatives. One kind of white blood cell disposes of old red cells when they are past their prime. But most are on the hunt for sources of disease and other unwanted substances that may have got into the body.

Although you can't make out individual white blood cells with the naked eye, you've probably seen a collection of one kind of white blood cell that have done their job and died – they make up pus. There is a whole army of these cells in your body, billions of them, each dedicated to taking on particular forms of attacker or internal cells that are in need of culling.

That's not the end of blood's armoury. There is also a third type of cell in the blood that may be less familiar – platelets. These are short-lived, rather shapeless cells that are responsible for blood clotting, preventing wounds from bleeding indefinitely.

The special molecule

Of course there's another component of blood as well – water. The plasma (we are not talking about states of matter now, remember) that the blood cells float in has a number of proteins and other chemicals dissolved in it, but it's primarily water. Your body contains lots of water – more water, in fact, than anything else. Water is a simple but fascinating molecule. One oxygen plus two hydrogens makes that most familiar of chemical formulae, H_2O. Water has huge significance for biology, so much so that when we search the Solar System for likely sites for life, we first look for water. Bacterial life has been found at the extremes of heat, cold and airlessness that our planet can serve up. But there is no known life without water.

Underlying water's importance is a unique collection of properties. It's the only compound that exists as solid, liquid and gas at the typical temperatures we experience on the Earth's surface. And as a molecule it has some surprising characteristics – without one of these, its boiling point would be below −70 degrees Celsius. If that were the case there would be no liquid water on Earth, and so no life. But thanks to this special property the water molecule shares with a few others, it boils at the familiar 100 degrees Celsius.

The property in question is hydrogen bonding, an attraction between the electrical charge on a hydrogen atom and that on another atom like oxygen, nitrogen or fluorine. In the case of water, the hydrogen's relative positive charge is attracted to the slight negative charge on the oxygen in another water molecule. The result of this bonding is that it's harder to separate the molecules into a gas than it otherwise would be. The bond has to be overcome, pushing up water's boiling point and so making the Earth habitable.

Hydrogen bonding is also responsible for another of water's unusual properties. Most substances occupy less volume as a solid than they do as liquid. However, solid water – ice as we tend to call it – has a higher volume than the liquid form, which is why it's not recommended to freeze a bottle full of water, and why ice floats on a pond, making it easier for life to survive under it. It's often said that this is a unique property of water. It's not – acetic acid and silicon, for example, are both less dense as a solid than as a liquid – but it is unusual.

Fill a small plastic bottle with water right up to the top, leaving no air, and screw the top on. Leave it in a freezer overnight. As the water expands to form ice it will either crack the plastic, force the top off or stretch the plastic so it feels strangely floppy once it has thawed. Don't use a glass bottle or you may get shattered glass all over your freezer.

The reason for this expansion on freezing is that the shape of the standard crystal form of water, a six-sided lattice, won't fit with the way the hydrogen bonds pull the hydrogen of one water molecule towards the oxygen of another. To slot into the structure, these bonds have to stretch and twist, pulling the water molecules further apart than they are at water's most dense form (which is at around 4 degrees Celsius).

Water is, of course, transparent but it does have a slight blue coloration due to the scattering of light (the same reason the sky is blue), although this is not obvious except when there's a large amount of water we can see through, for example in glacier ice.

One of the reasons water is so important for life is that it is a great solvent thanks to the charges on the molecule than make hydrogen bonding possible, dissolving many other materials and acting as a transport for them in living cells. But this isn't the only way that water supports life. It takes part in many of the chemical reactions necessary for the metabolic processes of the body. Without water, living cells can't exist.

A company of tiny boxes

I've already used the term 'cell' repeatedly. You can't avoid it once you start to take a look inside your body. The word was coined by Newton's contemporary (and arch rival) Robert Hooke. A great scientist in his own right, Hooke's best-known book is *Micrographia*, a wonderfully illustrated study of the very small, seen through magnifying glasses and early microscropes.

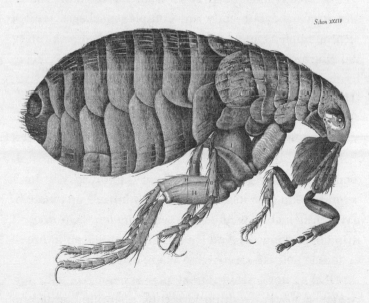

The illustration of a flea in Robert Hooke's *Micrographia*

Some of the illustrations folded out of the book, stunning the readers of the day with detailed images of a flea and a louse, two creatures with which they would be all too familiar, but which they would never have seen in such monstrous detail. He also amazed his public with a detailed drawing of the compound eyes of a fly. He even studied sections of cork. In these he saw an 'infinite company of tiny boxes' which he likened to the cells occupied by monks in a monastery. The biological cell is named after a monk's bedroom.

Every known living thing has at least one cell. The simplest forms of life – bacteria, for example – consist of a single cell, while your body has trillions of them. In effect each cell is a container of life. The blood cells we've already met are fairly unusual, but the more standard forms in your body are complex packages with a central nucleus and various bits of biological machinery floating around in the fluid surrounding it.

The superstar molecule

That nucleus houses the most famous complex chemical compound in existence, DNA. Let's face it, DNA is a celebrity of the chemical world. How many other molecules regularly get mentioned on the news? We don't even have to give its full name – the initials are enough. (Which is just as well as deoxyribonucleic acid doesn't trip off the tongue.) And we only have to see a picture of a double helix to know what we're dealing with.

DNA is not a single substance. It's not like salt, say, which is always sodium chloride, a simple compound of two atoms stuck together in the NaCl molecule.

DNA is more of a format for storing information in chemical form. The DNA in the nucleus of one of your cells – let's say a cell in the flesh of your fingertip where that blood oozed from – is in the form of a series of long molecules, twisted around proteins called histones that act rather like a set of spindles for the DNA.

You may have seen pictures of human chromosomes. Each chromosome is a single molecule of DNA with its accompanying histones, and each of your cells contains 46 of these chromosomes in the nucleus. We'll find out more about these in Chapter 7, but the important thing that often isn't mentioned when people are talking about chromosomes is that the DNA in each one is a single molecule. This isn't obvious because they are wrapped up in a bundle, making them much more chunky than a typical molecule. Their sheer size is also part of what distracts us from thinking of them as a molecule. The DNA in human chromosome 1 is the largest molecule known, with around 10 billion atoms in it.

Experiment – DNA dabbling

Here's a chance to experience what those forensic science dramas on TV are up to when they isolate a DNA sample. In this experiment you can extract DNA from a banana. It is the most complex experiment in the book, but even if you don't do it, it's still impressive that you can get hold of DNA with a relatively simple bit of science.

Blend half a banana to a paste (just blend for a few seconds – don't let it become too liquid). Mix clear liquid dishwasher detergent and a pinch of salt with around nine times as much warm water to fill half a mug (say 10cc of detergent, making up 100cc of solution). Stir this and the banana together, trying to avoid creating bubbles, until you have an even mix with no lumpy bits.

Use a coffee filter to filter the liquid from this mix in a cold place. Put some of the liquid in a narrow glass container (a test tube would be ideal) so it's a couple of centimetres deep. Now gently pour very cold alcohol down the side of the container so it forms a layer on top. DNA will begin to come out of solution in the alcohol. You should be able to spool it out on a cocktail stick.

Ideally the alcohol should be 95 per cent ethanol – effectively pure alcohol. If you can't get hold of this, rubbing alcohol should work. Alcoholic drinks are not pure enough. You don't have to use a banana – practically anything living will do, but bananas are one of the easiest things to use. Note that the final gunk will have some proteins attached, but it's mostly DNA.

That double helix structure of DNA is very similar to a spiral staircase. The helix part consists of long strings of sugars – the 'deoxyribo' in the full name of DNA comes from the sugar deoxyribose that forms part of these backbone polymers, long chains of atoms with a repeating structure. As far as DNA is concerned, these are just foundations. The important constituents are the treads of the spiral staircase. Each tread is made up of a pair of chemical compounds, which are selected from the four 'bases': cytosine, guanine, adenine and thymine.

Your own special code

These bases are like the zeroes and ones in binary code in a computer (though of course bases are not binary because there are four of them). There are six billion base pairs in the DNA that is found in each of your cells. The codes there are used to store information that will be used to produce various proteins, the multi-purpose workers of the biological world, and to create a whole set of other molecules that help determine how you are formed and develop over time. What makes the whole thing work is that the treads always have the same coupling of bases. Adenine is always paired with thymine, while cytosine is always linked to guanine.

This pairing is the key to a copying mechanism. New cells are produced by splitting one cell into two, and each of the resultant cells needs its own copy of the DNA data. To do this, the two chains of the double helix are unwound, dividing each of the treads in two. Although those two halves are not identical, because the base pairs always couple up the same way, it's easy to recreate the missing half and end up with a complete set of DNA in each cell.

DNA is often described as providing the blueprint for the living thing that contains it – and it certainly has quite a job to do. Just think of it. You started off as a single cell. That cell divided into two, the two cells divided into four and so on until you reached your current, magnificent, total of about 50 to 70 trillion cells. Clearly things couldn't just continue that way with simple splitting or you would just be a big blob of cells. Something had to give directions for the cells to know how to 'differentiate' – to form different types of cells and different structures – and that's the role of DNA.

However, to call DNA a blueprint is misleading. A blueprint gives you detailed specifications of exactly what goes where so you can build an artefact. But DNA has nowhere near enough data in it to specify everything that goes into a human being. There is certainly no link between the number of genes – the basic code level of the information in the DNA – a living thing possesses and its complexity. Rice, for example, has more than twice as many genes as human beings do. However, this is a simplistic view, as we'll see when we examine genes in a bit more detail.

Instead, then, it's better to think of the DNA in your fingertip (and every other normal cell in your body) as the control software of the complex automated factory that is a living thing. The DNA doesn't contain all the details, and other factors are interacting with the software, changing which parts of it are active at any one time. Nonetheless, as we'll see in more detail in Chapter 7, DNA has a hugely important part to play.

The 46 molecules of DNA in the nucleus of a cell aren't the only DNA in that cell, though. In fact there's some extra DNA that you could think of as alien – it doesn't originate in a human being at all.

The invaders in your cells

Floating around in a cell but outside the nucleus you will find structures called mitochondria. These minuscule pods are sometimes called the cell's power plants, as their job is to take the oxygen collected by breathing (delivered by the red blood cells) and combine it with chemicals from your food to make ATP, adenosine triphosphate, a molecule that your body uses to store up energy. The mitrochondria are biochemical battery chargers. The most remarkable thing about them is that they appear to have once been bacteria that became part of the cell in a mutually beneficial symbiosis.

This theory for the origin of mitochondria has been around a while, but the evidence for it became even stronger in 2011, when a common marine bacterium with the rather boring name SAR11 was discovered to be likely to share a common ancestor with our mitochondria. It's a bit like humans and gorillas – we both share a common ancestor and so, it seems, do SAR11s and mitochondria. Comparison of the genes in the two suggests that they originated in the same early form of bacterium.

This comparison was possible because mitochondria have their own DNA – just thirteen genes that are separate from your main chromosomes in the nucleus of the cell. Unlike your principle body of DNA, which is a mix-and-match combination from both your parents, the mitochondrial DNA only comes from your mother. These built-in ex-bacteria need the action of around 1,000 genes to work. In the distant past all those genes would have been on board the single cell that became a mitochondrion, but over time all but the thirteen have migrated out to the chromosomes.

The number of mitochondria present varies from cell type to cell type. They are at their most dense in your liver cells, where you will typically have over 1,000 mitochondria in each cell. Although mitochondria have a number of other functions, their biggest role is storing energy away in ATP, which is the chemical equivalent of a coiled spring in a clockwork motor.

When a spring is wound up, it takes energy to twist it into a tight form. That energy is stored until the spring is released, when it can push on a mechanism and make it go. Similarly, the mitochondria store energy by creating ATP. This rather messy chemical (its full name is dihydroxyoxolan-2-yl methyl (hydroxyphosphono-oxyphosphoryl) hydrogen phosphate) contains a pair of bonds that link phosphorus atoms with a single oxygen atom. These bonds (linkages between the electrons in the atoms) are relatively weak, and a simple chemical reaction will result in the bonds breaking, giving off energy in the process. It is the combination of tiny doses of energy from these molecules that gets your muscles moving every time you lift a finger or carry out any other action. Just to keep your eyes following this text, ATP bonds are popping all over the place.

Wearing your alien genes

Mitochondria aren't the only invaders that have been completely integrated into your body. Your DNA includes the genes from at least eight retroviruses. These are a kind of virus that makes use of the cell's mechanisms for coding DNA to take over a cell. (Aids is produced by such a virus.) These viral genes in your DNA now perform important functions in reproduction, yet they are entirely alien to human DNA.

If mitochondria were once bacteria, they are now very much part of your cells. Although they don't turn up in the more basic single-celled creatures, they are present in almost all organisms that have a nucleus in their cells. It seems the mitochondrial invasion took place at a very early stage of the development of more complex life on Earth. However, they aren't the only bacterial presence in your body.

Your trillions of tiny stowaways

Next time you take a look in the mirror, remember this. On sheer count of cells, there is more bacterial life inside you than there is human life. There are almost ten trillion of your own cells in that body – but as many as ten times more bacteria than that.

Many of the bacteria that call you 'home' are friendly, in the sense that they don't do any harm. Some are positively beneficial. They aren't as integrated into your system as mitochondria, so it is possible to live without them, but losing them makes life harder. Back in the late 1920s an American engineer decided to investigate whether animals could live without any bacteria whatsoever, hoping that a bacteria-free world would be a healthy one. James 'Art' Reyniers made it his lifes work to produce environments where guinea pigs and other animals could be raised bacteria-free from birth.

The result was clear: it was possible. You could clean away all those nasty bacteria and it wouldn't stop animals from living. As a bacteria-free world would clearly reduce the potential for disease, Reyniers' results encouraged the widespread use of antibacterial cleansers and antibiotics.

There is no doubt that some bacteria cause a huge amount of harm. It turns out, though, that Reyniers' research was misleading. He did indeed get some of his guinea pigs to live without bacteria. But many died. And those that did live had to be fed on special food. This is because bacteria in the gut help with digestion. This is particularly important for animals and insects eating plants high in cellulose, like grasses and wood. These foods are difficult to break down, and without bacteria to help, animals with this kind of diet wouldn't survive.

You could live without your bacteria – but without the help of the enzymes in your gut that bacteria produce, you would need to eat food much more loaded with nutrients than your usual diet. This is particularly true for vegetarians, as plant fibres are particularly resistant to our own enzymes and it's only with the help of the much wider range of chemicals produced by bacteria that we can get anywhere with them.

This is something you need to bear in mind if you take a course of antibiotics. Although any particular antibiotic will only kill a percentage of bacteria, there is no distinguishing between 'good' bacteria and 'bad' bacteria. Antibiotics don't care. They will, without doubt, cut a swathe through the bacteria in your gut. This means that you may need a richer diet for a while, and will also have to be careful to avoid infection – the bacteria in your gut help fend off unwanted intruders, so if you knock out these locals with antibiotics it is significantly easier for a new and possibly harmful strain to take hold.

Sadly for those who enjoy them, there is no evidence that adding 'friendly bacteria' in the form of pro-biotic drinks and other products has any positive effect. The bacteria consumed this way will make very little contribution to your in-house fauna. There is probably some psychological benefit (see page 267 on the placebo effect), but no genuine biological assistance from those friendly bacteria.

A useful appendix

Bacteria are also part of the story in what is probably the most misunderstood part of your body: your appendix. If you still have your appendix, you might wonder what the point of it is. After all, the appendix sometimes goes wrong and causes potentially life-threatening appendicitis, yet it doesn't seem to do anything useful. This surely does not make evolutionary sense. Given that human beings have had appendixes for a long time, if they are totally useless, why haven't they disappeared entirely?

It is only relatively recently that it has been discovered that the appendix is very useful to your onboard bacteria. They use it as a kind of holiday home; somewhere to get a respite from the strain of the frenzied activity of the gut; somewhere to breed and help keep the gut's bacterial inhabitants topped up. So the appendix isn't as useless as it has traditionally been regarded.

But it seems strange that the bacteria inside you, even those in the appendix, aren't mopped up by your defensive systems. White blood cells are constantly producing antibodies, proteins designed to lock onto invaders and cripple them. This is why transplant surgery is so difficult – human bodies even tend to fight off other perfectly harmless human cells. Yet by mechanisms we don't entirely understand, all these bacteria seem to be able to resist the actions of the antibodies.

One other surprise about the appendix is the recent discovery that it contains vast quantities of antibodies. Some of these do have a way of latching onto some of the bacteria that find their way to your gut, but in a helpful, rather than destructive way. The most common antibody in the gut, also very common in the appendix, is called IgA. This binds onto the gut bacteria – but not to kill them. Instead, it forms a supportive structure that helps the bacteria stick in place and thrive in the gut, rather than being flushed out as if they were food. Your antibodies give a helping hand to these useful gut bacteria.

The name IgA is short for immunoglobulin A. There are huge numbers of such proteins, large complex molecules produced in the body and used as chemical workhorses. Initially these were given sober and serious names like immunoglobulin, but over time a tradition has developed of landing them with quirky ones. So we have proteins called sonic hedgehog, pokemon, seahorse seashell party, dickkopf, R2D2, Homer Simpson, glass-bottomed boat and, my favourite, abstinence by mutual consent.

Bacteria don't know the five-second rule

Bacteria (and viruses) aren't, of course, always good for you. Although some illnesses are genetic or due to normal human processes going wrong, most are probably caused by one of these types of tiny invader. An old wives' tale that we need to check against our knowledge of bacteria is the five-second rule – the idea that if you drop a piece of food, as long as you pick it up within five seconds you should be okay.

Apparently this approach dates all the way back to the time of Ghengis Khan, though back then, when people were less fussy about what they ate, it was the twelve-hour rule. A US high school student, on a summer course at a local university, took a more modern scientific approach to the rule, with some interesting conclusions.

When Jillian Clarke took swabs from floors at the university, including areas with a high footfall, she discovered that the floors were surprisingly clear of bacteria. The PhD students helping her couldn't even find countable numbers of them. However, perhaps not surprisingly, they did discover that people are less likely to pick up from the floor and eat broccoli or cauliflower than sweets or biscuits.

Perhaps the most important finding was that when a surface was inoculated with E. coli bacteria, foodstuffs did pick up the bacteria in under five seconds – so in that sense the rule fails.

Worming their way into your affection

Bacteria may be the most common alien life form that you will have on and in your body, but they certainly aren't the only ones. Some people will have undesirable guests. Lice, for example (see page 17), or fleas, not to mention worms. Worms are fascinating – we tend to think of them purely as unwanted parasites, but there is now some evidence that the right worms in the right circumstances can be beneficial.

This may seem a bizarre suggestion, but though they are a more recent companion than the bacteria we depend on, human beings have lived with worms for sufficiently long that our bodies have grown used to them. Although trials are still relatively infrequent (quite possibly because of the revulsion worms cause), there is reasonably good evidence that some worms can have a beneficial effect on the body, because our internal systems expect them to be present and are out of kilter without them. It has been suggested that some medical conditions that have increased in frequency as worms have been wiped out could be improved with judicious application of worm therapy.

The noble leech

Another parasite that has a positive side is the leech. Leeches have been used medicinally for hundreds of years, but the traditional use was based on a totally false premise. Medicine has only recently become scientific. For a long time it hung onto an idea that was the medical equivalent of the Ancient Greek four elements, that of the four 'humours'. This was based on the belief that the body contained four liquids that maintained its equilibrium: blood, phlegm, black bile and yellow bile.

These humours had to be kept in balance. If you were thought to have too much blood, for example (and so were 'sanguine'), some would be removed by bleeding. This bloodletting was a common treatment, and often made patients significantly weaker and less able to fight off infection than they would otherwise have been. While it was frequently performed directly by incision, leeches were sometimes used as a convenient way to remove blood.

Although, thankfully, modern medicine has realised the ineffectiveness of bloodletting, leeches have come back on the scene to help with some post-operative problems. A blood-sucking creature like a leech wants blood to flow smoothly without clotting. To help this, it applies a natural anticoagulant as it sucks. An operation can sometimes result in congestion where blood builds up in some regions and doesn't reach others. Careful use of leeches can clear the congestion and help the blood to flow better into the tissues that are not receiving a good supply.

Aliens in the eyelashes

Depending on how old you are, it's also pretty likely that you have some other aliens on board. There are tiny creatures called eyelash mites that live on old skin cells and the natural oil (sebum) that is produced by human hair follicles. Unlike lice, these mites are only surface feeders and don't do any damage, though they can cause an allergic reaction in a minority of people. They are very small – typically around ⅓ of a millimetre when fully grown and near-transparent – so you are very unlikely to see them with the naked eye.

Put an eyelash hair or eyebrow hair under the microscope, though, and you may well find these little creatures, which spend most of their time right at the base of the hair where it meets the skin. Around half the population have them, with children having fewer and older people more. Although they don't have the positive benefits of bacteria, there is no need to worry about eyelash mites – they are harmless.

Seeing small

Such miniature invaders have only really become part of our conscious understanding of the body with the use of microscopes. Similarly, cells only began to be understood as this technology became more widely available. The first observations, like many of Hooke's, were done with a strong single lens, supported to avoid vibration. This was also true of the man who discovered bacteria in 1674, Anton von Leeuwenhoek. But real advances depended on the introduction of the compound microscope.

By simply putting two of the right lenses together in a tube, our ability to delve into the nature of microscopic life was much enhanced. A lens close to the object being studied produces a magnified image on the opposite side of the lens. This is a 'virtual' image – you can't see it, it floats in space. The second lens, the eyepiece, then acts as a magnifying glass focused on this already enlarged image.

We can thank a Dutch father-and-son team, Hans and Zacharias Janssen, for this invention. These Dutch spectacle makers put together their first compound microscope around 1590. At the time Hans was only a boy. He tends to be the better known of the two because his future career was based on optical instruments, but it's arguable that Zacharias should have most of the glory.

Our current knowledge of the working of the body has been enhanced greatly by other technologies that enable us to see beyond the immediately obvious. The first real breakthrough was the use of autopsies to explore the inner workings of the body, a process that was hampered because for many years it was illegal to undertake such operations. But cutting a person apart to see what's going on inside has its limitations, particularly if they are alive, and modern technology has a number of other answers to this need.

The rays that don't stop giving

The first big breakthrough was back in 1895, an accidental discovery when German scientist Wilhelm Röntgen was experimenting with a 'Crookes tube'. This was a crude form of the cathode-ray tube which was used in TV sets and computer monitors until LCDs and plasma took over. The 'cathode rays' of this tube are actually a stream of electrons, which can be steered using electrical and magnetic fields. The electrons usually end up hitting a phosphorescent screen which lights up where they arrive.

These glowing screens were built into the front of TV sets, but Röntgen had a free-standing screen, which he had left to the side of the tube rather than placing it at the target end. He was amazed to discover that it still glowed when he switched the tube on, despite the sides of his tube being swathed in cardboard to stop stray emissions. It seemed that the electrons, hitting a metal target, were generating some new kind of ray that shot off sideways and was so powerful that it went straight through the cardboard.

Röntgen referred to this new form of radiation as X-Strahlen (pronounced Eeks-Shtrahlen), which in

English became X-rays. The 'X' just meant this was something unknown and mysterious, and the term was only intended as a temporary nickname. The scientific establishment didn't like it and tried to call the effect Röntgen rays, but it was too late, the term 'X-ray' stuck.

Back then, just as now, a scientific paper sometimes caught the attention of the press, and Röntgen's paper on the discovery of X-rays had one feature that made the headlines: a single photograph. Röntgen had shone the X-rays onto his wife's hand. They passed through flesh, but not through bone. For the first time ever, the photograph showed a human skeleton inside the flesh; a picture of his wife's bones. It was even more striking as his wife had not taken off her wedding ring (although she seems to have tried to, as it's above the knuckle), so this stands out as a dramatic blob on the image.

The medical applications of this were so stunningly obvious that the world's first X-ray unit was set up at Glasgow Royal Infirmary in 1896, just one year after their discovery. The users of medical X-rays have never looked back. What's more, the general public could not get enough of the novelty of X-ray vision. Well into the twentieth century amateur electrical magazines featured DIY designs to build your own X-ray machine, and as a child my shoes were still being checked with a device that let you look down and see your own toe bones inside the shoe.

What was not realised initially was that, marvellous though X-rays are, they come with risks attached. Röntgen suspected from the beginning that they were a form of light, which they proved to be. X-rays are exactly the same stuff as visible light, but with higher energy. We know that electrons can be bumped up to a higher level by absorbing a photon, a quantum of light energy. But X-rays are so energetic that they can blast electrons right off the atom – they are what's known as ionising radiation.

Of itself, ionisation is a very common process. It happens, for instance, when salt is dissolved in water – so the fluids in your body contain plenty of ions. But when ionising radiation hits cells in the body it can create free radicals; highly reactive molecules that increase the risk of cancer. (The body's natural defence against free radicals is antioxidants, which is why foods with antioxidants in are often advertised as good for your health, though all the evidence is that antioxidants you consume don't join forces with your internally produced ones, so have no benefit.)

The danger of ionisation in your body created by the high energy photons means that it's best to avoid excessive exposure to X-rays, which is why radiographers operate from behind a protective screen. But the levels we are exposed to as patients are very low-risk, especially bearing in mind the natural radiation we are exposed to all the time. There is always a certain amount of radiation in the air around us from natural sources. A chest X-ray, for example, is about the same level of radiation as the extra natural radiation you are exposed to by taking a ten-hour flight.

Cats and nuclear resonance

To discover what is going on inside your body without cutting it open, doctors now have a much wider range of penetrating beams available to them. A CAT scan is still an X-ray, but one that goes far beyond anything that was possible before computers. It stands for 'computer assisted tomography' (or computerised axial tomography), which sounds a little scary when you realise that tomography is generally a matter of cutting things into very thin slices. But here it's the X-ray image that produces a series of snapshot slices through the part of the body being examined. Heavy-duty maths (hence the 'computer' part of the name) transforms data from a range of angles into a detailed, multi-layered image.

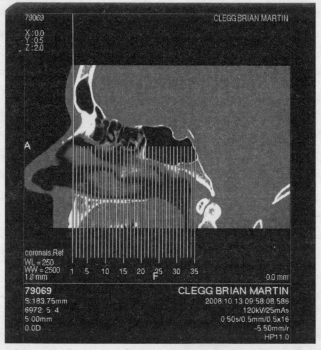

Image from a CAT scan performed on the author: vertical lines show the 'slices' used in other images in the series

The other well-known scanner is MRI, standing for magnetic resonance imaging. It was originally called NMR, with the 'N' short for nuclear, but that first initial was dropped because of the association of 'nuclear' with nuclear radiation. This was an unnecessary fear, as the name simply means that the nuclei of atoms in the scanned person's body are being observed. The patients aren't bombarded with radiation.

The protons in the nuclei of atoms can act like little magnets. MRI uses a strong magnetic field to get the magnetic fields of some of the protons in water molecules to line up. The scanner then uses a burst of radio. Radio is a relatively low-energy form of light, and if the radio photons have just the right energy they can give the little proton magnets a brief flip of the direction of their spin. The flipped protons rapidly fall back and produce their own photons, which can be detected. Because different types of tissue and different levels of blood flow produce different outputs it is possible to distinguish between them when the emitted photons are detected by the scanner.

Hunting the elusive neutrino

Photons of light of appropriate energies aren't the only particles that can pass through solid matter. Every second about 50 trillion particles called neutrinos pass through your body. These particles are emitted by the Sun and other nuclear sources. Neutrinos are very slippery customers. They are so difficult to detect that although theory predicted their existence in the 1930s, neutrinos weren't actually spotted for over twenty years. In an experiment at CERN in Geneva in 2011, these particles were thought to be discovered travelling faster than light, with claims that Einstein's theory of relativity would fall apart if something could do this.

Because of the ease with which they pass through your body, it might seem neutrinos would be great for medical scans – the trouble is that *no* part of your body is much of a barrier. Neutrinos have little more problem getting through you than empty space. In fact most neutrinos pass through the whole Earth as if it wasn't there. The only reason we can detect them at all is that just occasionally one of them will collide with an atom or molecule and will generate a little spray of other particles – we never see the neutrinos themselves.

Neutrino 'telescopes' are usually situated in mines a couple of miles underground, where hardly anything else is likely to get through and set off reactions in the vats of cleaning fluid, or similar materials, that are used as detectors. Such a device has been used to produce a neutrino picture of the Sun. It's very blocky – just a few pixels – and it's typical of neutrinos that the Sun was the opposite side of the Earth at the time.

The most dramatic neutrino detector is the IceCube observatory at the South Pole. This remarkable device, completed in April 2011, uses a square kilometre of ice as its detection medium, with detectors buried nearly 2.5 kilometres down looking for tiny flashes where incoming neutrinos collide with the ice above. The ice acts as both the barrier to other particles causing false signals and as a detection medium – there's something rather spooky about the thought of tiny flashes deep in the Antarctic ice revealing neutrinos from distant nuclear reactions in space.

The neutrinos light couldn't catch

The CERN discovery will probably prove to be a storm in a teacup. The experiment involved sending neutrinos down a distance of 732 kilometres (this incidentally has nothing to do with CERN's most famous experiment, the Large Hadron Collider). At the end of the journey, the few neutrinos that would be detected were discovered to have arrived 0.00000006 seconds earlier than they should have. By far the most likely reason for this is that the distance measurement was wrong. At the time of writing this result had not been duplicated elsewhere.

Failing that, the next most likely explanation is that the neutrinos were bending the rules. It's wrong to suggest, as so many articles did at the time, that modern physics somehow depends on nothing being able to go faster than light. Special relativity says that this won't happen as a rule, but it is possible to get around the 'barrier'. In fact we already have well-established experiments in which particles travel faster than light speed.

This is a consequence of quantum mechanical tunnelling. One of the strange aspects of quantum physics is that particles don't have an absolute location, just a probability of being in various places. This means that particles can jump through an obstacle without passing through the space in between.

This sounds like something obscure and unusual, but it's how the Sun (or any other star) works. For nuclear fusion to take place, positively charged protons have to be pushed incredibly close together – so close that even the temperatures and pressures in the Sun aren't enough to get the reaction going. The Sun only works because every second billions of particles tunnel through the barrier of the repulsion and fuse.

That same tunnelling technique has been used to send particles faster than light. All the evidence is that a tunnelling particle doesn't travel through the space in 'tunnels' through – instead it disappears at one side and instantly reappears at the other. So if you imagine a photon going 1 centimetre at the speed of light, tunnelling 1 centimetre instantly and going a further centimetre at the speed of light, it will have traversed the entire distance at one and half times the speed of light – 1.5c where 'c' is the speed of light.

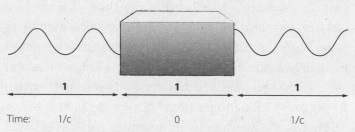

Diagram showing the action of a tunnelling photon

I'm not saying this is what is happening in the neutrino experiment, but I do imagine that the cause will be something similar. Not a collapse of special relativity, just a way around it. That's if it's not experimental error, which still seems most likely. Special relativity has been tested so many times and has always delivered.

Either way, neutrinos won't be joining the medical toolkit used to explore your body any time soon, but with the work of facilities like IceCube, they are of interest to astronomers. In exploring the universe, just as in investigating the innards of your body, it's light that reigns supreme. Light is our ultimate vehicle for exploring space, near and far, and it's one that your body is adept at handling.

4. Through fresh eyes

Your eyes are your most powerful mechanism for understanding the world around you – and their link to the rest of the universe is light. In this chapter we are going to discover just how much your eyes enable you to take in, and from how far away. Go out on a clear night and take a look at the sky. This may not be something you can do immediately, but do it when you get a chance. Take five minutes to really look up at the stars. If you have the time, take a chair out and look for a little longer. At first it may seem trivial, but it really is one of the most amazing experiences it is possible to have.

In Orion's belt

Let's say you can see the constellation Orion (it is visible pretty well around the world between November and February, is often visible at other times of year and is about the most easily recognised constellation).

Although constellations feature in a big way in astrology, they have no significance in science. They are, however, a useful way of picking out specific stars. Our brains understand the world through patterns. We're always looking for them – and we see them even when they don't exist. Constellations like Orion, the W of Cassiopeia or the distinctive Southern Cross, jump out at us because the pattern recognition modules in the brain find something they can latch on to.

Few people can see the images of the classical figures that most constellations are named after – Orion,

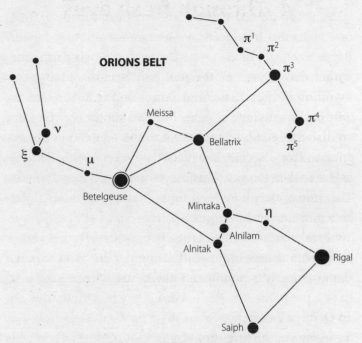

The constellation Orion

for instance, is supposed to be a hunter holding a club. But there is enough of a recognisable pattern in that collection of stars – particularly because of the straight-line proximity of the three stars in the hunter's 'belt' – for Orion to jump out of the sky at us.

Not only are constellations irrelevant to astronomy except as a pointer and name label, astronomy shows us just how much of an illusion they are. The stars in a constellation can be huge distances away from each other. The middle star of Orion's belt, for instance, is nearly twice as far away as most other stars in the constellation, but this isn't at all obvious.

Stars are named using a system introduced in 1603, in a star atlas produced by German astronomer Johann Bayer. Each star in the constellation has a two-part name, with a Greek letter as the first part, and the Latin genitive form of the constellation's name (the form meaning 'of' that constellation) as the second. In theory the stars are listed in order of brightness, but Bayer didn't always stick to this – so, for instance, the three stars in Orion's belt are Delta Orionis, Epsilon Orionis and Zeta Orionis. This doesn't work on brightness, but makes them alphabetical from north to south.

Stars that aren't in constellations usually get rather boring designations of letters and numbers. And to make things even more confusing, the better-known stars also have a pet name – a single word name by which they are more often referred to than their Bayer designation. So, for instance, the brightest star in Orion (the sixth brightest of all the stars in the sky, the bottom right star in the diagram of Orion), while technically Beta Orionis is better known as Rigel.

Similarly, the second brightest star in Orion, Alpha Orionis (the top left in the diagram), is more familiar as Betelgeuse. This too is in the stellar top ten and has a noticeable red tint. Betelgeuse is a huge star – a red supergiant. If the Sun were that big, it would stretch out nearly as far as Jupiter.

But if Orion is in sight, I want you to take a look at the middle star of the belt, Epsilon Orionis, known as Alnilam. It's time to give your eyes a workout.

If you've never really looked at the night sky, you might not have noticed that some stars (and at least one planet) have distinct colours. Next time there's a clear night, take a few minutes to stand outside and really examine the stars. After a while your eyes will become more sensitive. You should be able to pick out a few stars with a reddish tint and a few that seem a little more blue than the rest. If there's a very bright star that is very obviously red, it's probably not a star at all, but the planet Mars.

Alnilam is the most distant star in Orion, but as a bright-burning blue giant its distance doesn't particularly show. Alnilam is very young as stars go – only around four million years old (compare that with 4.5 billion years for the Sun). It is around 1,340 light years from the Earth.

Seeing into the past

As mentioned earlier, a light year is the distance light travels in a year, which given light's speed of around 300,000 kilometres per second is a fair range. Alnilam is around 12,686,155,200,000,000 kilometres away. Compare that with the furthest human beings have ever travelled, the distance to the Moon (a mere 385,000 kilometres) and you can see that we won't be visiting Alnilam any time soon. Yet without any technology, simply by opening your eyes and looking in the right direction, you can see an object that is 12,686,155,200,000,000 kilometres away. Your eyes are remarkable tools for exploration.

There's another strange thing about looking out at a constellation like Orion – it's a time jumble. Because light takes time to reach us, we see stars the way there were when the light set off, not the way they are now. Because all the main stars in Orion are different distances away, we see them at different times in the past. In the case of Alnilam, we are seeing it as it was around 1,340 years ago; in the seventh century. It's quite remarkable to think how much change has happened here on Earth while the light you see from Alnilam has been travelling towards us.

Waves or particles?

Let's take a moment to follow Alnilam's light from its creation to the moment your eyes detect it. Light is made up of tiny, insubstantial particles of energy called photons. You were probably told at school that light is a wave, and that is a useful way of looking at it, because photons have certain peculiarities that make them behave as if they were part of a wave. But that beam of light from Alnilam is still a stream of photons.

What is thought of as the wavelength or frequency of light when it is considered a wave is just the energy of the photons that make up the beam. This is what our eyes detect as colour and tells us where the photons come on the vast electromagnetic spectrum that stretches from radio waves and microwaves, up through visible light and into high-energy photons like X-rays and gamma rays.

The reason photons often seem to behave like a wave is that they have a property called their 'phase' that varies in a cycle with time. It's a bit like each photon having a little clock attached to it with a hand that sweeps very quickly around through 360 degrees. At any one moment in time, the photon's phase is pointing in a particular direction, and this corresponds to where a wave would be in its up and down wiggle.

Bursting from the heart of a star

The photons that reach your eye across space are created in the heart of a star as it undergoes nuclear fusion. In a star like the Sun, what's happening is that hydrogen nuclei – the tiny central part of hydrogen atoms, are being fused together to form nuclei of the next heaviest atom, helium. In the process a tiny amount of mass is lost and this mass is converted into energy, following the most famous equation in science, $E=mc^2$.

This equation tells us just how dramatic that production of energy is. The 'c' that is squared in the equation is the speed of light, so you get a huge amount of energy for a tiny amount of mass. This energy emerges in the form of photons (and other particles) inside the star. Almost immediately the photons will hit other particles and be absorbed, then further photons are re-emitted. This process happens again and again as the light gradually bounces its way towards the surface of the star. It can be a million years from the process starting to a photon emerging from the Sun.

In Alnilam, things are slightly different because it is burning so fast and furiously that all the hydrogen has probably gone, and it is busy producing other elements, but the effect is the same. After a series of emissions and absorptions in the depths of the star, eventually a photon will emerge from the surface. It will have much less energy by now after those billions of absorptions and re-emissions. Where initially it would have been far above the energy of visible light, and so be the type of light classified as a gamma ray, by now it will have dropped in energy enough to be visible – and it sets out into space.

The 1,340-year star trek

Once the photon escapes from the surface of the star, there is no stopping it without it being destroyed. Light has to travel at a specific speed or it can't exist. And so it flashes across space at 300,000 kilometres per second. The vast bulk of the photons that emerge from Alnilam will come nowhere near Earth. But a tiny few, including the photon we are following, will head in your direction.

For 1,340 years, through the last 1,340 years of our history, that photon will have been crossing space until it finally enters the Earth's atmosphere. If it's lucky it won't be absorbed by a molecule in the air. Many photons will. This is why a space telescope like the Hubble satellite can get so much better photographs than an Earth-based telescope. On the Earth, the air will always mean we lose some of the light. Although those air molecules will re-emit photons after they've absorbed them, they won't necessarily send them off in the same direction, so some of the light will be scattered into the sky, and some that continues in our direction will travel on a slightly different path, making the star appear to twinkle.

Finally, the photon arrives at your eye. This could be the exact same photon that left Alnilam 1,340 years ago. All that time it has been crossing space, only to wink out of existence as it hits your eye. If you wear glasses it will perish that bit sooner. As a photon moves through a substance like glass it is likely to be absorbed and re-emitted a number of times. And even if you don't wear glasses it won't be the same photon that you see, as that same process of absorbing and re-emitting will happen in the interior of your eye before a photon reaches your light detectors. Yet the process will be triggered by the photon that has crossed 1,340 light years of space from Alnilam.

The distorting lens

Eventually, a photon will hit the retina at the back of your eye. Along with many other photons triggered by originals from Alnilam, it will be concentrated on a small area of the retina by the focussing effect of your eye's lens. Like all optical devices, the lens depends on the way light changes direction when it passes from one substance to another to modify what is seen, a process know as refraction.

Experiment – The bending pencil

Fill a cup or a glass two thirds full with water (I find a straight-sided glass works best) and place a pencil in it so that it crosses from side to side of the glass, going all the way down to the bottom. Take a look at the pencil carefully at the point it enters the water. It looks as though it bends slightly, bringing it closer to going straight down into the cup or glass. It's not a huge deviation, but it is clearly noticeable that the pencil seems to change direction slightly. This is the result of the light bending as it goes into the water,

> just as it does (though even more so) when it travels from air to glass in a lens.

The traditional way of understanding this phenomenon responsible for the focussing of light in your eye is to observe that the light slows down as it goes into the glass of a lens (or the water in your cup). To keep the energy the same, this means the frequency has to go up – the waves come more often. If you imagine a wide beam of light hitting a piece of glass at an angle, the bit of the beam that hits the glass first will have an increase in frequency, while the light still travelling through air will maintain the same frequency. This will result in the wave bending.

Quantum theory's approach to light and matter is rather different. It says a photon will, in effect, take every possible path, with each path having a different probability. As a photon moves along a path, the property of the photon we have already met, called its phase, varies with time. Each of the different paths will give the photon a different phase at the point it enters the glass.

To find out what actually happens, you combine the phases of the different paths. Some will be opposite and cancel each other out. You are left with the phases that point pretty much in the same direction. And these cluster around the path that takes the photon the least amount of time. Although a particular photon can be thought of as following all of the potential paths, averaged out, the photon will be lazy and take the route that requires the least time. You might imagine this is the same as going the route that takes the least distance –

a straight line – but as your satnav often shows, it some-times pays to go a bit further on fast roads than taking the shortest route if it means dragging through the middle of a town.

The Baywatch principle

The way light behaves when it passes from air to water or air to glass is sometimes described as the Baywatch prin-ciple. Imagine there's a red-clad lifeguard on the beach who spots someone drowning. Their natural inclination might be to run straight towards the drowning person. But that isn't the quickest route. The best way is to run a bit further along the beach, if by doing so you can go a shorter distance through the water. Running on the beach is so much faster than running or swimming in water that it helps considerably to extend the journey on land a little, thereby getting to the person in trouble in the least time.

Exactly the same thing happens when light goes from air to a denser substance like glass (or water). Because the light goes slower in the glass, it will get to its destina-tion quicker if it travels a bit further through the air, then a shorter distance through the glass. The light takes the Baywatch route, arriving in the minimum time.

All this works on the assumption of light slowing down as it goes into glass, but light isn't easily slowed down. In fact it has to always go the same speed in any particular substance or it could no longer exist. But quantum theory explains why in fact it *does* slow down. Photons are always interacting with matter, specifically with the electrons on the outsides of atoms. When a photon comes close to an electron, the electron will eat up the photon's energy, becoming more energetic itself.

Usually, though, the electron isn't too stable in its new extra-energetic state. It easily drops back to its old state and sends out a new photon. That photon might head in the same direction, but it could head off in a totally different direction. Mostly, in a transparent substance, the re-emitted photons continue in the same direction, passing through the glass (or whatever the substance is) in a straight line. But they aren't going to get through as quickly if they spend time being absorbed and re-emitted, which they inevitably do. So the light slows down.

In an opaque substance the photon comes back out in another direction, away from the one it arrived in. It is from these new photons arriving at our eyes that we are able to see the object. It used to be thought that the light bounced off the object to get to our eyes, like a ball bouncing off a wall, but it really gets absorbed and re-emitted. Most objects are better at permanently eating up some colours (converting the energy to heat) than others. Depending on what colours of light the object absorbs totally and which it re-emits, we will see the object as a particular colour. For example, if an object absorbs all the colours of the rainbow except red, we will see it as a red object.

Looking through a lentil

Because of the shape of a lens like the one in your eye – roughly that of a lentil, which is where the word 'lens' comes from – the photons that spray out from a point are brought back to a point on the other side. The curved shape of the lens means that photons that hit it at different angles are bent by the angles necessary to bring them all together again. In the case of the lens at the front of your eye this process, focusing, produces an image of the distant object on the retina, which is how you are able to see.

There's only one problem with using a lens – they aren't very good at handling a range of colours. The amount a beam of light is bent depends on its colour. This is how a prism produces a rainbow from white light. With a traditional convex (bulging out) lens, blue light will be bent a bit more that the rest, red light a bit less. The result is that an image seen through a basic lens will have rainbow fringes distorting it.

The usual solution to this is either to have multiple lenses in a compound setup, where a concave lens helps correct the problems caused by a convex lens, or to use a mirror instead of a lens. Mirrors can also focus rays of light from different locations to a point, but they don't differentiate between colours. This is, in part, why astronomical telescopes nearly always use mirrors instead of lenses to collect their light. (A reflecting telescope is also much shorter for the same amount of power compared with a lens-based telescope.)

Through a glass, darkly

Just as with an opaque object, reflection of light off a mirror really isn't at all like a ball bouncing off a wall, once we understand it at the quantum level. When a photon hits a mirror it could reflect off at any old angle. (I use 'reflect' as shorthand here. Bear in mind that photons don't bounce off at all – each photon is absorbed by the mirror and a new photon is re-emitted. But the effect is as if it were reflected.)

Imagine a beam of light hitting a mirror and bouncing to your eye. Quantum theory says it doesn't have to travel to the middle of the mirror and reflect to your eye at the same angle like those optics diagrams most of us did at school. The photons have probabilities of taking every possible path, hitting anywhere on the mirror, then bouncing up at totally different angles to reach the eye. Each photon has a property called its phase which varies with time. If you add together the probabilities of taking the different routes, and the phase the photon would have along that route, most cancel out. The final outcome is that the light travels along the path that takes the least time – which usually happens to mean reflection at equal angles.

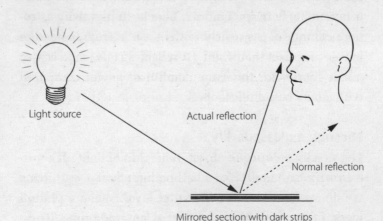

Light source   Actual reflection

Normal reflection

Mirrored section with dark strips

The action of light on a mirror with dark strips on it

But just because all those other probabilities are cancelling each other out doesn't mean they don't exist. And you can prove this. If you chop off most of the mirror, leaving only a section on one side of the centre in place, you obviously won't get a reflection from the missing middle. But put a series of thin dark strips on the remaining segment, designed to only leave available those paths whose phases add together, and it begins to reflect, even though the light is now heading off in a totally inappropriate direction for reflection as we understand it.

You can actually experience reflection happening at a crazy angle because of quantum effects without fiddling around with mirrors and dark strips. Visible white light is a mix of different colours of light, each of which will be reflected at a different angle by such an off-position mirror with dark strips on it. Shine a white light onto such a special mirror and you should see rainbows. Practically everyone has another sort of mirror that does the same thing – a CD or DVD. Turn one over to see the shiny playing side and tilt it against the light. The rainbow patterns you see are due to rows of pits in the surface cutting out all the paths with certain probabilities, leaving the different colours of light reflecting at unexpected angles into your eye.

The messy colours of sight

Mirrors may be great for focusing light without splitting it into its constituent colours, but your eyes wouldn't work if they had mirrors instead of lenses. Mirrors are no use for directing light from Alnilam (or anywhere else) into your eye. So the eye is left using a lens, and that means there will be chromatic aberration. If you actually saw what was produced by the lens in your eye, the picture would have colour distortion, leaving messy fringes around the objects you see. But as we will discover, the brain constructs the best image it can from the incoming data, and that process includes removing the chromatic aberration effects.

This means that with clever use of different colours on a piece of art it is possible to make it look three-dimensional or produce an effect that is uncomfortable on the eye. Red lettering on a blue background, for instance, can feel quite unpleasant to look at. A powerful contrast like this makes the chromatic aberration really stand out, and your brain can't cope with hiding the effects as it usually does.

Experiment – The lenses of your eyes

You can see a good example of what it's like when your brain simply can't edit out the chromatic aberration because it is too strong at the *The Universe Inside You* website **www.universeinsideyou.com**. Select *Experiments* and click on *The lenses of your eyes.* Take a look at the two versions of the word 'Illusion'. It's hard to put your finger on what is wrong with the image, but it causes a degree of discomfort as your brain tries its best to handle the extreme visual aberration.

Incidentally, one thing you have to bear in mind when understanding what your eyes are up to is that you can't see light. This seems a crazy statement. But what I mean is you can't see light the same way you can see a tree or a dog. Light hitting your optic nerves causes the sensation of sight. We see things when they emit or reflect light and those photons hit our eyes. But you can't see light as it passes by, because light doesn't bounce off other photons of light.

It's just as well. The space around you is filled with an inter-penetrating web of light and other forms of electromagnetic radiation that are invisible. Sunlight, artificial light, radio, TV, mobile phone signals, wireless networks – they are all the same stuff, and if they did bounce off each other then we wouldn't be able to use them – or to see. If you shine a powerful light down a black tube and look through a cut-away side, you won't see anything – the light going past the hole is invisible. It's only if there's something in the tube that scatters the light away from its path, such as the smoke used in laser displays, that you can see a beam passing by.

Picking up the photons

At the back of each of your eyes is a retina – a very special screen on the inside of your eyeball. It is onto these that the image of Alnilam is projected when you look up at the night sky. That screen is covered in an array of around 130 million tiny sensors which come in two forms, rods and cones. The rods just handle black and white. There are about 120 million of these, and they are significantly more sensitive than the three types of cone, which handle colour. When light is low, the cones give up entirely. In low light conditions, we see the world in black and white – something many people, children and adults, just won't believe until you demonstrate it.

If you have any doubt about the way your eyes switch off their ability to handle colour in low light, go into a room with good blackout curtains, or wait until night time and close your ordinary ones. Sit for a minute or two while your eyes get used to the light level. If it is not possible to see at all, put a torch under bed covers or a cushion so just a tiny amount of light creeps out. Nothing should be clearly illuminated.

Now look around you. Look at your clothes, your skin, objects around you. Even if it doesn't quite seem like a black and white movie, you will be unable to tell the colours of the things around you. If you can tell what colour they are, there is too much light – cut down the levels until you can hardly see at all and try again.

Ordinary colour vision works using the combination of the three primary colours, red, blue and green, from which any other colour can be created. You may have been told that the primary colours are blue, red and yellow, but this is simply wrong. These are simplified, children's versions of the *secondary* colours, cyan, magenta and yellow, which are visual negatives of the primaries. The secondary colours are the key colours for pigments – because pigments absorb the primary colours of light – but they aren't the true primaries.

Night vision is quite different from colour vision, registering only levels of brightness. But there is a crossover zone (called mesopic vision) when both types of vision occur together. When you experience this it's as if a whole new colour has been added to the spectrum that didn't exist before. Sight at this in-between light level has strange qualities – this may well explain why so many ghosts and other visual phenomena are seen at dusk. It's the time when our eyes are best able to mislead us, because two systems are competing to produce information for your brain to handle.

The colour-detecting cones are concentrated around the middle of the eye – if the light is very weak, you can see things better if you don't look directly at them, using the abundance of rods at the edges of your vision. Your eyes seem to be set up this way so that you can keep an eye out for predators creeping up on you at night. The three types of cones could be said to handle red, blue and green respectively, though the range of colours they handle actually overlaps strongly. It's more that their peak sensitivity is in a particular colour range. Not all animals have the same set of sensors. Some are colour-blind. Others, like dogs, have limited colour vision, with just two types of cones.

From light to mind

The photon that we have traced from Alnilam to your eye makes its way to the back of the retina (strangely the receptors in the eye are back to front, with the sensitive bits at the back, quite possibly as an accident of evolution). On the surface of each sensor are a set of special 'photoreceptor' molecules. When electrons in these absorb the photons of light, a tiny electrical charge is generated that is the starting point for getting a signal to your brain.

Some of the signals are combined at this stage before they are sent off up your optic nerve. There are considerably fewer fibres in the nerve than there are sensors, so there has to be some pre-processing in your eye before the signal gets to your brain. Mostly the connections from your right eye go to the left side of your brain and those from your left eye to the right side of your brain, but a proportion of the fibres cross over to the other side, so some signals from your right eye are handled alongside the left eye information. This crossover is to enable 3D vision to work – in birds, for instance, where the eyes work more independently than ours do, there is much less crossover.

At this stage what we have is a series of electrical signals. The brain now processes these using a collection of modules that handle different aspects of vision. These modules (not separate parts of the brain, but separate functions within it) deal with motion detection, the selection of detail, pattern recognition, shape recognition and so on.

After this initial processing, your brain ends up with a set of data, which it uses to build its picture of what you see. It constructs a night sky with the star Alnilam currently at the centre of your focus. This is totally different from the way a camera takes an image. What you 'see' is an artificial construction the brain makes from all those signals and processing. In a way it's much less 'real' than a simple photograph.

Your artificial view of the world

The artificial nature of sight is why optical illusions work. Your brain is always constructing images of the way it *thinks* things should be, rather than the way they are optically. The picture projected on your retina, for example, is upside down – the brain turns it over. This trickery by the brain can be graphically demonstrated by wearing special glasses that flip your vision upside down. After a few hours the brain has had enough of being messed about and turns the image the right way up. Someone wearing these inverting glasses starts seeing things properly again.

Experiment – Confusing your brain

Here is a simple example of the way that your brain's sophisticated technology for identifying shapes and shading can be used to produce a misleading image.

The chessboard optical illusion

We're all familiar with the layout of a chessboard, and our brains know how to process the effects of shadows. But this image is drawn in a way that specifically misleads the interpretation of those effects. It seems quite clear in the image above that square A, one of the black squares, is much darker than square B, one of the white squares. In fact, though, they are both exactly the same shade of grey.

You may find this hard to believe, but you can prove it if you fold the page and bring the two squares together. You will discover that they are exactly the same shade. If you don't want to mangle the book, go to **www.universeinsideyou.com**, click on *Experiments* and then the *Chessboard experiment*. This has a video where the square marked A is moved down to the square marked B, so that you can see they really are the same shade.

Another example of the brain's cheating is the way that it removes the blind spot. Part of your eye doesn't work. Where the optic nerve joins the retina there aren't any sensors. But your brain combines input from your two eyes to make the blind spot disappear. Similarly, when you are looking up at the night sky your vision seems steady and unmoving, but in reality your eyes regularly make little darting movements called saccades.

This fluttering around of the eyeballs helps your brain build a more detailed picture of the world around you. Saccades take place very quickly – they are the fastest of all external movements of parts of your body – sweeping the eyeball through around 10 degrees in as little as 1/100th of a second. If you saw a true representation of what your eyes took in, everything would be constantly blurred and jumping about, so the brain simply edits out the bits that you don't need to see.

Quantum reality

We've heard several times that the photon that crossed space to your eye allowing you to see the stars is a quantum particle, but what does that really mean? 'Quantum' is one of those words we hear often enough, but it's not always clear what people are talking about. It doesn't help that the word is used so loosely, whether it's in strange products that offer something like 'quantum vibrational therapy' or in the common usage 'a quantum leap' which seems to turn the meaning of 'quantum' on its head.

'Quantum' in the sense used by physics means the smallest amount of a particular something that can exist. It's a tiny packet of something. As we've seen, the term was used originally to refer to what would become known as photons, but now, in the sense of quantum particles and quantum physics, it's the science that deals with very small particles and their behaviour.

Once scientists became aware of the quantum world in the early part of the twentieth century, it didn't take long to discover that this is a strange, *Alice in Wonderland* place where particles do not behave like smaller versions of the larger objects we are familiar with in everyday life. When we throw a ball, we can predict what it is going to do exactly (given enough information). But when looking at a quantum particle, we can only give probabilities of where it is and how it moves. Until we actually make a measurement and pin the particle down, only the probability exists.

Through Young's slits

Probably the simplest example of quantum strangeness is an experiment that dates back to the early 1800s, called Young's slits. It was used to 'prove' that light was a wave. To carry out the experiment, a tight beam of light is sent through a pair of narrowly separated slits. The mingled beams from the two slits then fall on a screen at some distance behind. Instead of appearing as two bright blobs, one for each slit, the result is a series of light and dark fringes on the screen.

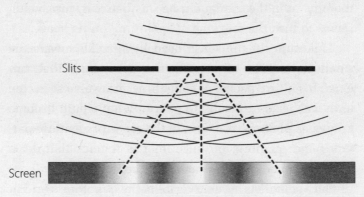

Young's slits

This was taken to show that light was a wave, because those fringes seem to be an interference pattern. When two water waves cross each other you can get a regular pattern set up. Where both waves go up at the same point you will get a strong upward undulation. Similarly at points where both waves go down, you will end up with a dip. But if one wave goes up at a point where the other wave goes down, they will cancel each other out, ending up with level water. This is interference. If light was doing the same thing, the dark fringes would be where the waves had cancelled each other out and the bright ones where waves had added together.

Such interference doesn't seem possible with particles. Imagine a large number of pieces of putty thrown at a wall through two slits – there would be no pattern of fringes built. And yet now we know that light *is* a stream of photons. So how do they achieve the effect? Amazingly, even if you fire one photon at a time at the pair of slits, eventually an interference pattern builds up. What could individual photons be interfering with to cause the fringes?

Here comes the quantum strangeness. This happens because each photon goes through *both* slits and interferes with itself! Remember that a quantum particle can be thought of as going along every possible path from A to B, each with different probabilities. Because it doesn't have an exact position, but rather is a combination of all these different possibilities, a single photon will go through both slits. The probability of where it can be found is spread out like a wave, and it is, in effect, this probability wave that causes an interference pattern for the particles.

If you put special detectors into the experiment which specify which slit the photon went through, but still let it pass through, the pattern disappears, producing a pair of bright blobs on the screen, just as you would expect with pieces of putty. If you make a measurement, forcing the photon to be in one place rather than a spread-out range of probabilities along all different possible paths, it can't go through both slits. It's enough to just look at a photon to cause it to totally change behaviour.

Uncertainty reigns

Quantum theory may seem obscure, but bear in mind every time you use you eyes and look at something, this is a quantum process at work. In fact, your whole body is made up of atoms, each of them made up of quantum particles. Probably the best-known term applying to quantum particles is the 'uncertainty principle'. This is sometimes interpreted as meaning that nothing is certain in a quantum universe – but it isn't that kind of philosophical concept. The uncertainty principle (sometimes called Heisenberg's uncertainty principle after the German scientist who devised it) simply states that the better you know one of a pair of linked pieces of information about a quantum particle, the less well you know the other. For example, the more accurately you know where a particle is, the less accurately you can know its momentum (that's its mass times its velocity). Know its momentum exactly and the particle could be anywhere in the universe.

A good way of picturing the uncertainty principle is to imagine taking a photograph of a particle. If you take the picture with a very quick shutter speed it freezes the particle in space. You get a good, clear image of what the particle looks like. But you can't tell anything about the way it is moving. It could be stationary; it could be hurtling past. If, on the other hand, you take a photograph with a very slow shutter speed, the particle will show up on the camera as an elongated blur. This won't tell you a lot about what the particle looks like – it's too messed up – but will give a clear indication of how fast it's moving. The trade-off between momentum and position is a little like this.

Getting entangled

There are many (many!) more mind-boggling happenings at the quantum level, but I just want to briefly mention the most remarkable, which is called quantum entanglement. This says that you can link together two quantum particles so that they effectively form a single entity, even though one could be triggering sight at your eye while the other is light years away in space. Often this link involves a particular characteristic of a particle, like its spin.

Quantum spin is a funny thing – it's not really about a particle spinning round like the Earth does. It's a measurement you can take of a particle that is digital. That means that when you measure it in any particular direction, it can only have one of two values, up or down. Before you take the measurement, the particle doesn't have a value for its spin, it just has a probability of the various different outcomes.

It could, for instance, have a 50:50 chance of being up or down. So half the time, making a measurement on such a particle, you would get the value 'up' and half the time you would get the value 'down'. Until you make the measurement, though, you have no way of knowing which you will find, because the particle isn't in one state or other – it's in what's known as a superposition of the states, both up and down at the same time, just like the photon goes along every possible path until you pin it down.

Now imagine we link together two such quantum particles. We can entangle them in such a way that when we measure the spin of one, we know for certain that the other one will have the opposite spin. (There are a variety of ways to do this. The simplest is to create two photons from the same electron at the same time.)

Now, here comes the clever bit. You can separate those two particles as far as you like – sending one to the opposite side of the universe if you wish – and when you check the spin on the 'home' particle and, say, it's up, you know for certain that the other one is down.

It might seem that this isn't such a big deal. After all, imagine you had a pound coin and sawed it in half along the narrow edge. You end up with two half-width coins, one with a head on it, one with a tail on it. You put one half coin, unseen into your pocket and send the other half off to the opposite side of the universe, again, without looking at it first. Now look at the half in your pocket. It's heads, so instantly you know that the other half is tails. It's not rocket science. But the quantum particles are totally different from this.

The half coins had the value of 'head' or 'tail' from the moment you made them. But when you make the entangled particles, neither of them has a value for spin pre-defined. Each is genuinely both up and down with a 50 per cent probability of being either when measured.

The two particles are identical. It is only when you look at one and it randomly settles into the up position that the other, instantly, however far the distance, becomes down. A message has crossed the universe instantly. It's possible to test to see whether the particles already have the information secretly hidden away or come up with it when they are looked at – and there are no secret values.

If you could use such a mechanism to send a message it would reach anywhere in the universe instantaneously. In practice, though, there's no way to send useful information this way. The results sent down this spooky link are random, so can't carry anything meaningful. You can't choose if the spin is going to be up or down, it happens by chance.

Even so, the way entanglement transfers information can be used for some remarkable applications, from ways to keep data securely encrypted and computers that can solve problems that would take conventional machines the lifetime of the universe to solve, to quantum teleportation – a miniature version of a *Star Trek* transporter that makes it possible to create an exact copy of a particle or collection of particles at a remote location.

A normal whole from quantum parts

Perhaps the ultimate paradox of quantum theory is the existence of your body. As we have seen, every bit of it is made of quantum particles – every atom inside you is a collection of quantum particles. Your senses operate on electrical and chemical impulses that are processes involving quantum particles. When you see that light coming from the distant star Anilam, it is a quantum particle that has crossed space, and a quantum process than enables your eye to detect it.

Your body is a quantum machine, and yet you see and experience a normal, apparently non-quantum world where probability doesn't reign, and things can't be in more than one place at a time. I wish I could provide an explanation for this – but I can't. No one, from the most exalted physics professor down, understands why the quantum building blocks of reality behave one way, while our everyday experience is totally different. At the moment all we can do is shrug and say 'That's the way it is.'

A galactic feat

Let's look back at that night sky. If you are in the northern hemisphere there's one other feature that it's worth taking a look at in your exploration of the universe through your body's capabilities. One of the most recognisable constellations is Cassiopeia. Again pattern recognition is at work here – the five main stars of the constellation form a large letter W, which is hard to miss (though you may see it looking more like an M).

But it's not Cassiopeia itself we are interested in.

If you think of Cassiopeia as a W, treat the second V in the W as an arrow and follow its pointer by a distance that is about the same as the entire span of Cassiopeia. This will have taken you into the much less obvious constellation called Andromeda. And around the point you arrived, a little fuzzy patch of light is just visible with the naked eye. If you were to see it through a good enough pair of binoculars it would become obvious that this isn't a normal star.

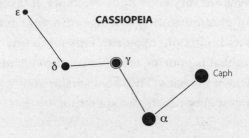

The location of the galaxy Andromeda

If you can see that little patch, you are seeing as far as is humanly possible without magnification. Your eye is undertaking an amazing feat. That fuzzy smear is the Andromeda galaxy, the nearest large galaxy to our own Milky Way. But 'near' is a relative thing in intergalactic terms. The Andromeda galaxy is 2.5 million light years away. When the photons of light that hit your eye began their journey, there were no human beings – we were yet to evolve. You are seeing an almost inconceivable distance.

Your eyes are very good light detectors. It takes only a handful of photons to trigger a signal in your brain. Yet sight has its limitations. You can only see a tiny portion of the light that is pouring towards you from Andromeda and elsewhere in space. Those sensors in your eyes only react to a very tiny part of the spectrum.

Glow-in-the-dark urine

The range of vision extends a little further in other animals. Many birds, for example, have an extra set of cones that stretch into the ultraviolet. This comes in handy for the hawks you see hovering high above the roadside, hunting for small mammals. They're on the lookout for mice, voles and shrews, which are pretty well camouflaged against wild grass in ordinary light. But these little animals urinate constantly – and their urine glows in ultraviolet. The hawk doesn't so much spot its prey as follow a trail of wee and pounce.

There is a way you too can see ultraviolet, if only indirectly. When you look at a fluorescent object it seems almost to glow of its own accord. Usually when we see something, the photons it re-emits are in the same energy range as those it absorbs. But fluorescence involves an object absorbing ultraviolet photons, then giving off visible light. So you see 'extra' light coming off the object as a result of incoming radiation that was originally invisible. The same thing happens with fluorescent light bulbs – ultraviolet is produced inside the bulb and that stimulates a fluorescent outer coating to give off visible light.

Experiment – Fluorescence in action

Get hold of an ultraviolet light source. You can buy ultraviolet lamps quite cheaply, but if you have a flat-screen TV that glows blue when there is no signal, this is also a good source of ultraviolet. Try a series of potential sources of fluorescence. Look for objects with 'dayglo' colours. Try a white shirt that has been recently washed – whites detergent has added material that fluoresces to give a 'whiter-than-white' tinge. You will also find that the more garish magazine covers and product packaging are often fluorescent to grab the eye.

Ultraviolet and visible represent only a fraction of the light spectrum. As you stand in your garden looking up at the stars, you are bathed in a whole range of photons that your eyes are unable to detect. Least energetic is radio, from broadcast stations to WiFi and mobile phones. Then there are microwaves, used for shorter range communication, as well as radar and the eponymous ovens. And just before we get to visible light, there's infrared, which you can feel as heat.

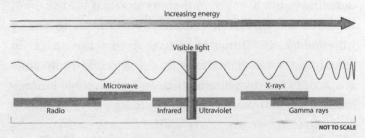

The electromagnetic spectrum

Finally, even more energetic than ultraviolet are X-rays and gamma rays. The distinction between the two reflects how they are produced. X-rays are produced the way ordinary light is, from electrons on the outside of atoms giving off energy. Gamma rays come out of the nucleus of an atom. There's considerable overlap in energy between the two. They are both called 'rays' for historical reasons – but they are exactly the same photons as any other part of the spectrum, just with higher energy.

Remnants of the Big Bang?

All these different kinds of photons, including the visible light picked up by your eyes, are streaming towards you from the stars. The further they come, the further back in time you see. The photons that have been on their way longest are sometimes called the echo of the Big Bang, and for a very good reason – they seem to come from everywhere and nowhere.

Although televisions with manual tuning that pick up an analogue signal (rather than a digital one) are comparatively rare nowadays, you have probably seen one. If you have, you will be familiar with the snow of white dots that dance around on the screen when the set is not tuned in to a particular channel. Some of this is earthly interference, but some is coming from outer space. In actual fact a television like this is a crude radio telescope, picking up photons that set out on their journey around 300,000 years after the Big Bang – over 13 billion years ago.

You've also probably seen radio telescopes, at least in photographs. They are usually big dishes, some of them hundreds of metres across. These dishes act like a mirror in an optical telescope, collecting together the radio signals from some distant source and focusing them on a receiver. But for the television to pick up those photons in the way we just mentioned, it doesn't need to have its aerial pointed in the direction of the Big Bang. This raises an important question: if the universe started at a single point, as the Big Bang theory says it did, where was that point?

Hold up one finger, approximately 30 centimetres in front of your nose. Take a finger on your other hand and hold it so the end of it is very near the tip of the first finger. The point between those fingers is the place where the Big Bang happened.

This seems a ridiculous statement. How can I possibly know that you are standing in just the right place to identify where the Big Bang happened?

The expanding universe

To explain this, we need to explore another strange thing about the universe. If you look out at distant galaxies, they are almost all moving away from our own. With the exception of a few really near galaxies like Andromeda (which is *really* near on the scale of the universe, at just 2.5 million light years!) everything is heading away from us. It seems amazing that we just happen to be located at the centre of the universe, and hence where the Big Bang happened. Too amazing, in fact.

Experiment – Blowing up the universe

To understand why the Big Bang happened at that point in front of your nose, and why we appear to be at the centre of the universe, get yourself a balloon. Draw some spots on it with a felt pen – these represent galaxies. Blow the balloon up a bit and see how far apart the galaxies are from each other. Now blow it up a bit more and look again. How are the galaxies moving?

The spots representing galaxies all move away from each other. However, the spots aren't actually moving across the balloon. They are still on the same bit of rubber as they always were. Instead, it is the balloon itself that is getting bigger. Similarly, it is the *space* within the universe that is expanding. So wherever you are in the universe, all the other galaxies move away from yours, just as happened on the balloon, but no galaxy can claim to be at the centre of the universe.

Now let the air out of the balloon. It gets smaller and smaller. This is like running time backwards. In practice the balloon will stop shrinking when it gets back to its original size. but imagine it got smaller and smaller until it was a tiny dot. Every bit of rubber would be in the dot. You could choose any place on the balloon while it was still inflated and it would end up at that single point. In the same way, the Big Bang happened everywhere in the universe. Wherever you are you can say 'this is where the Big Bang happened,' because the entire universe is the location of the beginning of everything.

The reason some galaxies head towards us is that they are so close that gravity pulls them in our direction faster than the expansion of the universe takes them away. In about five billion years, Andromeda will plough into our own galaxy, the Milky Way, and after much disruption the two combined will form a super galaxy. In case you are worried about possible effects on the Earth, don't be: a) you won't be around, and b) the Earth will have already been crisped by an expanding, reddening Sun.

So the Big Bang happened everywhere around us – and that's why you didn't need a radio telescope to detect the echo of the Big Bang, or the cosmic background microwave radiation, as it is more formally called. It comes from everywhere. If your senses were able to detect microwaves, you would constantly see the glow of the early universe filling the sky – as it is, we can pick it up with the right kind of detectors.

We can't see all the way back to the Big Bang, because right at the beginning everything was so compact and energetic that light couldn't get through it. It was like trying to look through the Sun to the other side, but even more so. After around 300,000 years, though, things had cooled down enough for the universe to become transparent and hugely powerful gamma rays, light at its most energetic, started blasting across it.

All the time, the universe continued to expand, giving that light (which came from everywhere) more and more space to cross. One effect of expanding space is that the light reduces in energy. Imagine someone throws a heavy ball at you, then throws the same ball while they're running away at top speed. The second ball would hurt less because it would have less energy, having expended some crossing the extra distance. Similarly, the light from the expanding universe has less energy than it had when it was first emitted. And if photons have less energy they move down the spectrum.

Visible light moves towards the red (this is called a red shift) – and those gamma rays gradually shifted down through X-rays, ultraviolet, visible light, infrared and have ended up as microwaves. It's these microwaves that produced the images of the after-effects of the Big Bang, captured by satellites called COBE and WMAP, and that produce some of the fuzz on the TV screen.

The probable Big Bang

I need to put in a proviso here. The Big Bang theory is our best-supported current scientific theory of how the universe began, but it's not definite, and it's not the only theory considered by serious scientists. We are working with very indirect evidence, and not just because we can't see past that 300,000 year mark. All the evidence we *do* have supports the Big Bang theory, but it is not without its problems.

For example, the Big Bang theory says everything started out of nowhere and no time in a singularity, a point in space-time of infinite density and infinite temperature. When things go infinite, the equations that predict what is happening break down. The theory that the idea of the Big Bang is based on simply doesn't work any more at that point. So we cannot be absolutely sure that the Big Bang was the beginning of everything, as the maths used to make the predictions breaks precisely where it matters most.

There are other theories that get around the difficulties with the Big Bang's singularity, but they too have problems. For the moment, the Big Bang remains our best theory, and for that reason it tends to be referred to as if it were fact. But this isn't an experiment we can check in the lab, or even through direct observation of something in space; it is a conclusion from various indirect measurements and a whole lot of model building.

Playing with models

The models used are not actual physical models. Real models do sometimes get built in science – famously, when Crick and Watson worked out the structure of DNA, their first action was to build a stick and ball model of a section of DNA – but usually when scientists say they are building a model they mean a mathematical model. This is a set of rules and numbers that should give the same results as what's observed in the real world. As long as the model's predictions and reality agree, then we have a possible explanation of what is happening in the universe. But when the model's predictions and reality go adrift it's time for a new theory.

A good example of this is the discovery that galaxies behave badly. All that keeps the stars in a galaxy together is gravity, and there is an opposing force that is trying to split them up. Like pretty much everything else in the universe, galaxies rotate. If you looked out and spotted the Andromeda galaxy, all your eyes could detect is a faint pattern of light. Your body's capabilities are amazing, but sometimes we need technology to help, and with modern telescopes we can see enough detail to discover that galaxies are indeed spinning round. As they spin, the stars in them are trying to shoot off in a straight line. The only thing that stops them is gravitational force, pulling towards the centre of the galaxy.

There's a catch, though, that shows there is something wrong with this model. If you calculate the mass of everything we expect to be in a typical galaxy and add it together, there's not enough mass to hold the galaxy together at the speed it is spinning. It should be spraying out stars like a demented pinwheel. There must be something more holding it together than the gravitational attraction of the matter we know about.

Of course not all matter in a galaxy is obvious. We can see the stars and glowing clouds of dust, but we can't make out planets or black holes or cold dust. But even allowing for all these there should be more. The most popular model to explain this phenomenon incorporates 'dark matter'. We don't know what exactly this might be (though there are some suggestions) but it's essentially extra mass that only interacts with the familiar stuff through gravity. It seems impervious to electromagnetism, and hence light.

This isn't the only possible model, though. An alternative is that gravity behaves subtly differently on the scale of galaxies. After all, we know that the universe operates very differently on the quantum level to the way we see ordinary-sized objects behaving. Perhaps galaxy-sized things have their own rules. This theory is called MOND, for modified Newtonian dynamics. It takes only a very small change in the effect of gravity to explain away that extra rotation speed.

The out-of-control universe

Another example of a model in action dealing with something we can't quite understand, is 'dark energy'. This is required to explain away something very strange about the expansion of the universe. You would expect that an expanding universe would gradually slow down. This is not because of friction, the reason things tend to slow down in the familiar world, but because of gravity. All the various bits of the universe are pulled towards each other by gravity. This gravitational force acts as a brake on the expansion.

It was more than a little surprising, then, when it was discovered that the expansion of the universe seems to be accelerating! Apparently the universe is not just getting bigger, but the rate at which it is getting bigger is going up. If this is the case (it's just possible that there is another cause for the indirect measurements which have been interpreted as acceleration), then something must be driving the acceleration. It takes a lot of energy to get the universe's expansion to speed up, and this is what has been given the label 'dark energy'.

These two dark components account for most of the universe, which is totally mind-boggling, when you think about it. Remembering that matter and energy are interchangeable, we can say that around 70 per cent of the universe must be dark energy to keep the expansion accelerating at the rate it is. Around 25 per cent should be dark matter. That leaves just five per cent for all the

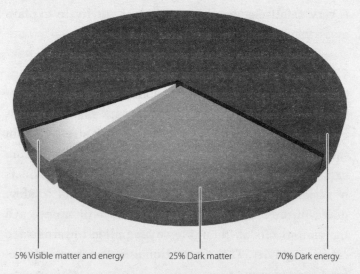

5% Visible matter and energy   25% Dark matter   70% Dark energy

Pie chart of the percentages showing how small
'ordinary stuff' is – probably unnecessary

matter (including your body) and light that we are familiar with. A remarkable 95 per cent of the content of the
universe is unknown!

This could be seen as rather depressing, in that it high-
lights how little science really understands, but I find it
delightful. We're not totally ignorant, after all – we know
vastly more about the nature of matter and light and the
universe than we did just 100 years ago. And yet there
is still so much more to find out! When Max Planck, the
man who devised the basic idea behind quantum theory,
was at university at the end of the nineteenth century
he had the choice of being a scientist or a musician. His
physics professor advised going into music, because
pretty well everything in science was now known. How
wrong that professor was.

A quasar too far

Staying on the subject of things we're not totally sure about, while the Andromeda galaxy is the most distant thing your eyes can detect, pretty well the most distant things we can detect in detail using telescopes are quasars. When they were first discovered it was thought that quasars (a neat shortening of 'quasi-stellar objects') were distant stars, but the colour spectrum of the light coming from them didn't seem right. It's too red.

As we've already seen, when objects in space move towards us, their light gets an increase in energy – it's blueshifted. And when they move away, their lower energy light is redshifted. The light from quasars is shifted a long way into the red. Because of the expansion of the universe over time, the further something is away, the greater its redshift. The first quasar to be studied, back in the 1960s, turned out to be (at the time) the most distant object ever observed. Yet its brightness was comparable with a star in our own galaxy.

After more study with better instruments it was discovered that quasars emit as much light as a whole galaxy, from an area that can be as small as our Solar System. Many have a pair of 'jets': very energetic streams of glowing material spurting out from either side. It seems likely that quasars are baby galaxies, still forming. Most galaxies are thought to have super massive black holes in their centres. With a mature galaxy, like our Milky Way, that black hole will have dragged in all the nearby debris, but in a young galaxy it will still be pulling in nearby material.

It's all this material, accelerated to near light speed as it plunges towards the black hole, that is thought to give off the quasar's dramatic blaze of light. As for the jets, a likely possibility is that the black hole has a sphere of material orbiting around it, spinning with the black hole and prevented from plunging into it by its spin. At the poles there will be no spin to speak of, leaving gaps through which material could be blasted. This explanation is very much at the speculative end of cosmology, though – there is no strong evidence to confirm it.

Black hole myths

While quasars remain fairly obscure, even though the name is probably familiar, I was able to mention black holes earlier without needing an introduction. Black holes have become part of the language, conjuring images of a bottomless pit that can swallow anything and that never lets anything go. Black holes have become an essential part of the mythology of the cosmos, featuring as the dark, all-consuming spirits of space.

Like most myths, though, you can't believe everything you've heard about black holes. Firstly, they may not even exist. Einstein's general relativity predicts that they *can* form, and we have very good indirect evidence for them, but in principle they might not be real. The evidence could be produced by some other phenomenon.

Then there's the idea they're a kind of universal vacuum cleaner, sucking up everything and anything that dares to come near. There's an element of truth in this picture, in the sense that all stars are good at clearing nearby space because they have a strong gravitational pull. But a black hole, which is formed when a star collapses, no longer able to sustain itself against its own massive gravity, only has the same gravitational pull as the star that formed it. (Don't worry, by the way. The Sun can't become a black hole; it isn't big enough.)

If you were in orbit around a star at the point in time that it collapsed into a black hole, you would continue to happily orbit it without being pulled in. But a black hole is much smaller than a star of the same mass. The black hole itself is theoretically of zero size, a 'point singularity' (though as with the Big Bang, what this really means is that the theory breaks down and we don't know what goes on). The black hole's apparent size is its 'event horizon', the sphere around it being much smaller than the original star which is the point of no return. Pass the event horizon and the gravitational pull is so strong that nothing, not even light, can get out.

Building a black hole

The radius of a typical star forming a black hole might be something like 1.5 million kilometres – but the event horizon for such a star once it collapsed into a singularity would just be fifteen kilometres in radius. Because you can get much closer than you can to a conventional star, the gravitational pull becomes much stronger – gravity goes up with the inverse square of distance, so halve the distance you are from a black hole and you quadruple the gravitational pull. Objects pulled towards the black hole will get up to sizeable percentages of light speed as they get close to the event horizon.

A black hole also gives a whole new meaning to tide marks. Tides are simply forces caused by the differing gravitational pulls at different points in space. As you approached a black hole you would experience a dramatic tidal force. Your body would become the ultimate gravitational experiment.

Imagine being in a space suit, heading for the black hole feet first. Your feet would feel a much stronger attraction than your head. The difference in pull across the length of your body – the tidal force – would stretch you so much that you would end up like a long, thin piece of pink spaghetti. This process is known as spaghettification (despite rumours, scientists do sometimes have a sense of humour).

This deadly stretching would not necessarily happen before you reached the event horizon, though. You could still be alive at that point – how soon spaghettification kicks in depends on the size of the black hole. A very big black hole, like those thought to be at the centre of galaxies, would have a very gentle increase in gravity. You would slip past the event horizon without noticing it. But you would still be stretched to a string as you headed towards the centre of the hole, that is if you survived the bombardment of radiation produced by fast-moving debris on its way to the centre.

I've said that the centre of a black hole, called a singularity, is in theory a point. But that hides one last really weird thing about black holes. The singularity, technically, is not a point in space, it is a point in time. General relativity, the theory that predicts the existence of black holes, says that gravity is a warp in space *and* time. At the heart of a black hole time itself is well and truly twisted. Once you pass through the event horizon you are headed for a point in time, not a point in space. The time of your total obliteration is fixed at that moment.

Black holes and quasars are amongst the most exotic inhabitants of the universe, but there are more familiar aspects too, many of them pumping photons in your direction as you look into the night sky and triggering a response from the detectors in your eyes. We've already met galaxies, vast collections of stars that can have anything from a few billion to 100 trillion stars inside them, and we think that there are around 150 billion galaxies in the universe. It's a big place.

Our own galaxy, the Milky Way, home to around 300 billion stars, can be seen on a really dark night as a faint band across the blackness of space, but the really obvious inhabitants of the night sky are relatively local stars and, nearest of all, our own Solar System. With the naked eye you can see five planets – Mercury, Venus, Mars, Jupiter and Saturn, with Venus and Jupiter being the two brightest things in the night sky after the Moon. But all those photons reaching us from planets have had a double journey. They only arrive at Earth after first setting off from our Solar System's prime light source, the Sun.

The non-eternal sunshine

It's when you take a look at the Sun (not literally – it will damage your eyes, even when partly obscured) that you can really see how amazing those billions upon billions of stars out in the universe are. The Sun is a nice enough star, but it's nothing special. Pretty much average in size and power. It's middle-aged, too – at around 4.5 billion years old, it is around halfway through its lifespan.

The light from the Sun is, to all intents and purposes, white. White light isn't really a colour, it's simply the whole bag of visible colours thrown together. Yet when someone draws the Sun they usually make it yellow. And when you see it at sunset, dimmed enough that your eyes don't naturally avoid it, our neighbourhood star looks red. It might seem that we're a bit confused, but this is down to photons of light, busily interacting with matter again.

In this case, the matter is the air. Many of the photons that enter the atmosphere from the Sun just hurtle straight through, but a fair number will be absorbed by gas molecules in the air, then re-emitted. When they are re-emitted in a new direction it is called scattering. This process is selective; the more blue the light, the more it gets scattered. This is why the sky is blue during the daytime – because that blue light is being scattered away from the Sun's position more than the colours towards the red end of the spectrum.

If sunlight contained equal amounts of all the colours, the sky would be violet, the most scattered of all the colours we can see, but there is considerably more blue than violet present, so that dominates. With some of the blue photons pulled out of the initially white light, what's left has a yellowish tinge, so that is our usual perception of the Sun. And when the sunlight has to go through significantly more atmosphere, as it does when the Sun is setting and the rays are going tangentially across the planet, the dominant colour left coming directly from the Sun is red, so we see that dramatic red sunset.

The Sun may be an average kind of a star, but it's anything but average as an inhabitant of the Solar System.

It's 1.4 million kilometres across, over 100 times the size of the Earth, and weighs in at one third of a million times greater in mass. Over 99 per cent of *everything* in the Solar System by mass is in the Sun. And, as everyone knows, it's hot. The surface is a relatively chilly 5,500°C, but at its core it is closer to 10,000,000°C.

The power source of life

If we are to use your body to explore science, it's important to realize that it wouldn't exist or be able to function without the light coming from the Sun. For a start, without it you wouldn't be able to see – but you owe far more to the Sun's light than that. Firstly it's where the Earth gets most of its heat from. A small amount of the Earth's heat comes from the planet's core, but the majority reaches us in the form of sunlight. Without this constant source of energy the Earth would be far too cold to live on.

What's more, you couldn't breathe or eat without the Sun. The oxygen you breathe comes from plants, which produce it as a by-product of photosynthesis. Light energy is used in photosynthesis to produce the chemicals (principally carbohydrates) that fuel life. Photosynthesis is much more complicated than the photoelectric effect used in solar panels, where light blasts electrons out of a special material to produce electricity. The chemical processes in photosynthesis are complex and often amazingly fast – some of the reactions are the fastest ever measured, taking place in under 1/1,000,000,000,000th of a second.

The light is absorbed in plants by pushing up the energy of electrons in special pigments like the chlorophyll that makes plants green. This *is* like the photoelectric effect, but there's more to it than that. The energy from the light is then transferred in chemical form to an in-plant reactor, the photosynthetic reaction centre, where a fundamental reaction that produces the oxygen as a by-product is performed. It's this oxygen that you breathe. Different plants have different levels of oxygen production, and despite all we hear about rainforests being the planet's lungs, it's actually plankton in the seas that make the greatest contribution to the atmosphere.

Animals like us don't share the plants' ability to convert light energy into food. We have to use an intermediary – either eating a plant, or another animal (which itself will have eaten a plant, or another animal, etc.). Indirectly, though, the power source of almost all life is the Sun.

Not only our heat, oxygen and food, but the majority of our usable energy comes indirectly from the Sun. Fossil fuels formed because the Sun powered the plants that would eventually form those deposits. Solar energy is obviously from the Sun, but so too is wind power, as the weather systems are powered by sunlight. The only exceptions are geothermal energy and nuclear power.

Is there anybody out there?

We need that energy to exist, as do all living things, and there's certainly plenty more energy out there in the universe to potentially support life. As you stand looking out at the stars on a dark night you are seeing many possible homes for other life. Our Sun is just one of billions of stars in our galaxy, and there are billions more galaxies. The chances are that there is life out there in the universe, but I wouldn't hold your breath until it is discovered.

The Solar System is not a very encouraging habitat. In the early days of science fiction, it was often imagined that there was life on the Moon, Venus or Mars. None of these is likely to support life. Venus is an overheated hellhole where lead runs liquid and clouds of sulphuric acid fill the sky. The Moon and Mars have limited water and atmosphere, and are very cold. While it's just possible that some sort of bacteria-like life could exist in a carefully protected pocket in one of these locations, it's unlikely. And the other planets are even less likely to support life.

The best chance for life in the Solar System outside Earth is one of the moons of Jupiter, Europa. At first sight this isn't a great location, far too far from the Sun to have the warmth needed to support life. The surface temperature on Europa is around −160°C. But Europa has a secret beneath its surface; under its icy crust it is likely there is liquid water, warmed by a combination of the huge tidal forces exerted on the moon by Jupiter and by its radioactive interior.

If Europa really does have this ocean with temperatures above freezing, it is possible, though not at all certain, that some basic form of life could have evolved there. Water and appropriate temperatures aren't the only factors, however. All the life we know of depends on carbon, and although some people have speculated that it would be possible to make living things out of silicon instead, that element isn't as flexible as carbon in the way it joins up to make large molecules – an essential quality for producing life. So there would have to be plenty of carbon and other atoms around too – but there is the possibility Europa could support life.

The intelligence test

All this is not to say there couldn't be plenty of intelligent life in the universe, but it is much more likely to exist on a planet orbiting a distant star. Despite the distances involved, we have now found hundreds of planets outside the Solar System. The first were spotted by the wobble that the star gets as a planet orbits it. This technique tends to pick up big planets like Jupiter, as they produce the most obvious wobble. Other methods have detected more Earth-like planets, smaller and probably rocky, not made of gas. But there is no evidence yet of life, and certainly not intelligent life.

Despite a lot of effort going into the search for extraterrestrial signals, nothing has been found. Earth has now been pumping out radio signals for around 100 years, so there is a 'mist' of radio signals 100 light years deep around us. In principle, anyone with the right technology in that radius could detect us. Of course, life forms within that distance might not be intelligent, and even if they were they might not use radio, but it is slightly disappointing that nothing has emerged yet on this front.

Even if we did spot another intelligent life form at a very close interstellar distance like twenty light years (the nearest star other than the Sun is four light years away, and twenty light years is still very much in our galactic backyard), we couldn't make much headway with a conversation. If we used radio to communicate – which as a form of light is the fastest thing around – we would have to wait 40 years to get a reply every time we asked a question (that's after working out *how* to communicate)!

As for visiting any alien civilizations, it's pretty well out of the question. We are seriously challenged by the technological difficulties of sending a human being to Mars, which is just four light *minutes* away on a good day. It's estimated it would take six months for a manned mission to reach Mars. The nearest star other than the Sun is more than half a million times further. Without some technology that allows us to bend the restrictions of light speed like a *Star Trek* warp drive – not technically impossible, but vastly beyond our foreseeable technical capabilities – we aren't going to visit the stars.

We are isolated, if not alone

The same goes for alien visitors. There have been plenty of legitimate UFOs – in the sense of being unidentified flying objects – though many have proved to be optical illusions or aircraft that were simply not identified. But any alien spacecraft have exactly the same problems with the distances involved that we have, and it is likely that all alien encounters have been hoaxes, self-deception or error.

Even the term 'flying saucer' is controversial. It was first used in a newspaper report in 1947 to describe the sighting of some unusual craft by US pilot Kenneth Arnold. At the time Arnold did not say that the vessels he saw were shaped like saucers. Instead he commented that they moved erratically, 'like a saucer if you skip it across a pond.' The word was picked up by newspaper headline writers and then misunderstood as the shape of the craft he saw. Soon after, sightings of saucer-shaped craft became common.

We may not be alone, but we are certainly fairly isolated here on Earth.

And yet we have followed photons from the far reaches of the universe, from quasars and distant galaxies, and from our life source, the Sun, to our home planet where some of them are detected by your eyes. It's time to come back down to Earth, quite literally. Perhaps after all that star gazing, your stomach is rumbling – your eyes may be directed to the stars, but your stomach has a much more earthly focus.

5. Marching on the stomach

When noises emerge from your stomach it could simply be time to eat, but it could also warn that indigestion's on the way. Not the most worrying problem your body can face, but still a matter for discomfort. So perhaps you reach for an indigestion tablet, which is not really a medicine, just a simple component in a chemical reaction.

Physics examines what atoms *are*, but chemistry gives us an understanding of how they combine. It's sometimes said that chemistry is all about electrons, because chemical reactions usually involve interaction between different elements when they share or exchange electrons from the outer layer of their atoms.

Your inner chemistry

Your stomach has some serious acid on board – hydrochloric acid. It's one of those acids that they take a lot of care with in the chemistry labs at school, because it is strong enough to inflict serious damage. But then that is just what is needed in your stomach. The job of that acid is to break down whatever you eat so that it can be used to generate energy, as well as making it easier to dispose of the waste.

The acid levels in the stomach vary and sometimes this can result in discomfort. So can the action of the stomach acid on various parts of your insides it shouldn't reach, for example when there is 'reflux', where the acid sprays out of the stomach and up into the oesophagus. These problems are most common as a result of bad eating habits (such as overeating or eating too late at night), although some people suffer from it as a result of physical problems like a hiatus hernia.

The instant remedy? Pop in an antacid tablet. The result is a simple chemical reaction.

Although there are various kinds of antacid on the market, many contain a carbonate like calcium carbonate or magnesium carbonate. The carbonate part is a carbon atom with three oxygens attached.

Reach for a chunk of rock

Calcium carbonate is a very common mineral. It gives eggshells their solidity and it is the main constituent of limestone, marble and chalk. Yes, you are, in effect, eating a powdered rock when you take an indigestion tablet, though I wouldn't recommend this as a cheap substitute.

Carbonates are great at reacting with acids. This is unfortunately obvious in areas that suffer from acid rain. Buildings constructed from marble, and particularly the softer limestone, have a bad time when the rain is acidic, eroding visibly. Sculptures lose definition, while inscriptions can disappear entirely, leaving whole graveyards with blank, silent markers.

But what's bad for stonework is excellent for your stomach.

When a chemical like calcium carbonate meets up with hydrochloric acid, there is a chemical reaction. A simple chemical reaction involves different bits of two compounds (molecules that contain more than one element) swapping places. They do this because of energy. There's energy in the way the different electron bonds join the atoms in molecules together, but not all bonds are the same. If there's energy given off when you go from one configuration to another, it's usually easy for that swap to take place. It's a bit like something dropping from a height. It's easy to get a rock from the top of a cliff to the bottom, because potential energy is lost as it falls. It's a lot harder to get a rock from the bottom of a cliff to the top, because you have to put energy in.

Experiment – Stomach basics

Drop an antacid tablet (a simple one, rather than the 'dual action' variety) into a glass and pour on a little vinegar. You should see a stream of bubbles coming from the tablet. This is the same reaction that takes place in your stomach when you take such a tablet – a reaction that releases carbon dioxide. Vinegar is a much weaker acid than hydrochloric, so the effect will be less dramatic. If nothing happens, it may be because the tablet is coated in a protective layer.

Repeat the experiment with a tablet broken up into bits. You should see a more vigorous action. This is partly because you are getting through any outer coating, but also because you have increased the surface area of carbonate exposed to the acid.

In the case of calcium carbonate and hydrochloric acid, the result is a vigorous interaction that ends up with three molecules being produced. Hydrochloric acid has one hydrogen and one chlorine atom bonded together.

In the reaction, the chlorine marries up with the calcium to form calcium chloride, while pairs of hydrogen atoms grab an oxygen atom from the carbonate to make water, and the remains of the carbonate become carbon dioxide gas. As a result acid levels drop and, hopefully, your stomach becomes less uncomfortable.

The evil compound of life

Carbon dioxide is a simple chemical compound that has a bad name these days. If it were in a Bond movie, it would be the evil villain intent on world domination. This notoriety is because of its role in global warming as a greenhouse gas. And it's certainly true that too much carbon dioxide in the atmosphere is not great. But it's important not to paint this gas in too bad a light, because there are a couple of reasons that you wouldn't be alive without it.

The first benefit carbon dioxide gives you is the desirable side of the greenhouse effect. In the atmosphere it acts like a kind of one-way mirror for heat. Most incoming sunlight shoots straight through the atmosphere. It warms up the surface of the Earth, which then gives off lower-energy infrared light. Some of this is absorbed briefly by the carbon dioxide molecules. They then re-emit the infrared, sending some onward into space, but some back to the Earth. So the carbon dioxide acts as a sort of blanket, insulating the planet and keeping it at a habitable temperature for us.

If you want to see what carbon dioxide can do at its worst, take a trip to Venus. This was once thought to be quite similar to Earth, but with a 97 per cent carbon dioxide atmosphere it has a runaway greenhouse effect. The average temperature is 480°C, and it can get as hot as 600°C. We only have around 0.039 per cent carbon dioxide by volume in our atmosphere, but the greenhouse effect as a whole (other gases like water vapour and methane contribute) means that Earth's average temperature is about 33 degrees higher than it otherwise would be. With no greenhouse effect, the Earth would have an average temperature of −18°C, severely limiting life.

The other essential role of carbon dioxide is feeding plants. As we saw earlier, the Earth's life cycle is founded on plants. Even carnivores need them, as they will eat animals which have to eat something as well – inevitably, somewhere down the chain you will come to plants. Plants take in carbon dioxide from the air, using the carbon to grow through photosynthesis and producing the oxygen we need to breathe as a waste product (see page 123).

Adding a little fizz

And then there's the fun side of carbon dioxide, discovered surprisingly early on. Scottish doctor Joseph Black was the first to isolate carbon dioxide in 1756. Just eleven years later, Joseph Priestley, who later discovered oxygen, started to study the carbon dioxide produced in the Leeds-based Jacques brewery. One of the tests Priestley did on the gas was to bubble it through water, where he found some dissolved, making ordinary water taste like the sparkling mineral water from the Alps.

Priestley pretty much forgot this until 1772, when he was dining at the Duke of Northumberland's home in London. As part of the entertainment, the guests were given distilled seawater to drink, which proved very bland. Priestley announced he had a way to improve it, and returned the next day to turn it into soda water. By then, Priestley was making carbon dioxide with sulphuric acid and chalk, a not dissimilar process to the antacid tablet dissolving in your stomach. He had been banned from the brewery after ruining a batch of beer when he tried to dissolve carbon dioxide in ether. Sadly, though, Priestley never got round to making soda water commercially, so it was easy pickings a few years later for the Swiss Johann Schweppe.

Sitting at Dmitri's table

At school, chemistry probably seemed to be dominated by the forbidding, ramshackle structure of the periodic table. Yet that strange table makes it possible to predict how the acid in your stomach will interact with an antacid tablet. Although it can be daunting, the periodic table was a huge breakthrough for science when Dmitri Mendeleev came up with it. The Russian scientist was not the first, nor the only, person to look for some kind of order in the various elements that make up the world around us. But he was certainly the most dedicated to the task, playing endlessly with a pack of cards, one element per card, looking for ways to arrange them that would make sense.

The principles of the periodic table are simple. It has a series of rows, each with more massive elements than the previous one, each getting heavier from left to right across the row. These rows are arranged in columns, where elements within a particular column will have

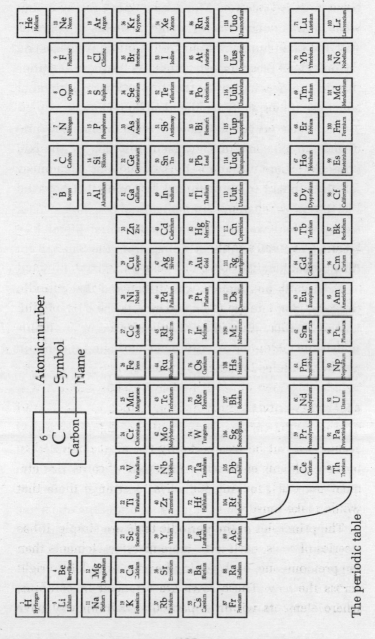

The periodic table

behaviours in common. Mendeleev didn't realize it, but what he was doing was arranging elements in columns with the same number of electrons (or empty spaces) on the outside of the atom's structure. As it's these electrons that determine what bonds the atom will form with other elements, they specify its chemical behaviour.

This idea proved its power when Mendeleev predicted the existence of new elements that no one had observed. There were gaps in the table, and Mendeleev felt these ought to be occupied by atoms that behaved like the known elements immediately above them. So, for example, there was gap under silicon, which Mendeleev labelled 'eka-silicon'. ('Eka' comes from the Sanskrit for the number 'one'.)

Eventually an element was discovered that fitted in that gap, later known as germanium. It has a lot of similarities to silicon (both would later be used in producing transistors and other electronic devices), and it behaved just as Mendeleev predicted it would.

Meet element 114

This use of the table to predict what a new element would be like has carried on up to the present day, though not every chemical is quite as predictable as germanium. Take element 114. At the time of writing this element doesn't have a real name. It just goes under the nickname ununquadium (literally 'one-one-four-ium' in Latin). At the moment the heaviest element with a name is 112, copernicium.

This is an element your stomach is never going to have to deal with. These ultra-heavy elements don't occur in nature. The heaviest thing around is uranium, element 92. Everything above this is made in nuclear reactors and particle accelerators. Such conditions are required for the creation of these elements because the force that holds the nucleus of an atom together, the 'strong force', has to work hard to overcome the repulsion between all the positively charged protons in the nucleus.

The strong force has one significant failing – it only works over a very, very small distance. So by the time atoms get to the size of uranium, with its 92 protons (the number of the element tells you how many protons in the nucleus and how many electrons there are around it), it is right at the limit of the strong force's ability to keep things together. Any larger and nuclei tend to be very unstable.

Most of the really massive elements only last for thousandths or millionths of a second before they split apart, but ununquadium occupies what's known as an 'island of stability'; a region of the table where atoms are more able to cling onto existence because the number of particles in the nucleus lets them get into a particularly stable form. The isotope of element 114 with atomic mass 289 stays around for seconds at a time.

Isotopes, as we've seen earlier in the book, are variants of elements that have different numbers of neutrons in the nucleus. The simplest atom of all, hydrogen, has a nucleus that is just a single proton. Add in a neutron and it still behaves as hydrogen does chemically, because it still only has a single electron, and it's the outer electrons that determine what an element will do in chemistry. However, with that extra neutron it is now heavier and behaves differently in nuclear reactions. We have gone from hydrogen to the isotope deuterium.

Because practically all the mass in an atom is in the nucleus, the atomic mass is just the number of protons plus the number of neutrons. So when we talk about the isotope of ununquadium with atomic mass 289, it has $289 - 114 = 175$ neutrons in its nucleus.

Element 114 was first produced in 1998 at the Joint Institute for Nuclear Research at Dubna in Russia. Just one atom of this isotope was produced in the first experiment, and though a range of isotopes have been produced since, it has only ever been made a few atoms at a time. Given how little there is, and that it only stays around for a few seconds, we have no idea what ununquadium looks like – it was expected to be a silver-grey metal like the majority of elements in the same region of the periodic table.

Heavy metal or noble gas?

The periodic table predicts that ununquadium should behave a bit like lead. In Mendeleev's terms it's eka-lead, the element below that metal in the table. Amazingly, though, despite element 114 having only been made a few atoms at a time, it's thought that it actually behaves more like a noble gas than a metal.

The noble gases are the least friendly column in the periodic table. Their outer set of electrons is full, meaning they don't show a lot of interest in reacting with other elements. These are gases like helium, neon and xenon. Apart from their uses in specialist lighting, the best known of these gases, helium, is something of an oddity. It was first discovered in the Sun, before being found on Earth. It is a hard-to-come-by element in the air, as it drifts off into the upper atmosphere before it's noticed. Remarkable, then, that we can buy a cylinder of it just to inflate balloons and get squeaky voices. (Most helium is extracted from natural gas deposits.)

With so little of it to study, how is it possible to say that ununquadium behaves more like a noble gas than a metal?

The atoms of the element are passed down a thin tube with an inner coating of gold. This tube is held at a temperature that varies constantly along its length, from room temperature at one end to a chilly −185°C at the other. As atoms pass down the tube, the reducing temperature means they have less and less energy – they don't bounce around as much.

The expectation is that metals like lead would bind onto the gold relatively easily and not get far down the tube. But the antisocial noble gases would get a lot further, before a weak attraction finally got them to stick in place. The element 114 atoms make it to the end of the tube, suggesting they are more noble than lead-like.

This isn't quite the failure of the periodic table it seems. It's likely that relativity is getting in the way of chemistry. Because there are a large number of electrons flying around the outside of the atom they need to go faster than usual. And special relativity (see page 198) tells us that the faster something moves, the more mass it has. The expectation is that these fast-moving electrons will have enough extra mass to change the way they move, altering chemical properties.

Turning food into energy

Whatever ununquadium is really like, your stomach is highly unlikely to encounter any, but it will certainly handle plenty of other atoms in your lifetime. Technically the stomach is just the pre-processor of the gastric system, though we tend to use the term to describe the whole business of turning food into energy. In the stomach, your food is attacked by the hydrochloric acid we've already met and by enzymes – complex chemicals that specialise in breaking down the proteins in food matter – until you've got a semi-digested mess that passes into the intestines.

The point of this assault on the food you have eaten is to gain access to relatively simple chemicals like sugars and fats that are made up of carbon, hydrogen and oxygen. Extra oxygen is brought into the system, carried from your lungs by your blood. This oxygen reacts with the sugars and fats, oxidising them. We've all experienced oxidation reactions before, when something burns – and that produces heat. In effect, the reaction in your body is a slow chemical burn, adding oxygen to produce carbon dioxide, water and energy. This energy is then stored away by mitochondria in chemical form (see page 56).

An unusual difference in the way we interact with our food compared with other animals is the way we prepare it. Not only do we wash what we are going to eat to remove impurities, we often go much further – we have a way of making food more digestible than it naturally is: cooking.

Hot food is good food

No one is sure just how cooked food became an important part of human life. It's generally assumed it started off by accident, when an animal or some grains fell close to a fire. The attractive smell may have encouraged passers-by to sample the chargrilled food, and the enhanced flavour that results from cooking would have made it something well worth copying.

One of the effects of heating food is to modify the texture of proteins, making them easier to chew and digest. Cooking also releases complex chemicals that stimulate our sense of smell. We tend to think that when we eat food it is taste that drives our likes and dislikes, but smell is a very strong component of our system for detecting what is good to eat. You don't want to have to taste excrement (for example) to know it's not going to make a good meal.

The sense of smell is the first line of defence in preventing us from eating something harmful, and much of what we think of as taste is actually smell. Some of the enhancement of taste in cooking is due to the breakdown of carbohydrates into simpler sugars and the concentration of flavours as water boils off, but much of it is due to the release of aromatic chemicals which stimulate the nose.

Before long though, a more important side effect of cooking was noticed – it also has the advantage of killing bacteria and viruses, and of destroying some toxins, such as phytohaemagglutinin, the poison in kidney beans (which are deadly when uncooked) and the poisons found in nightshade-related plants like potatoes.

It must have taken some time to realise that not only was cooked meat more tasty, and easier to eat because it was less tough, but also that those who ate it were less likely to suffer stomach pains or to die from their diet. But once that realisation was in place, we could add food to our diet that in its natural state would have been totally inedible.

This realisation must have been particularly difficult with foods that are poisonous when raw, like the kidney bean. It's hard to imagine anyone seeing a neighbour die as a result of eating kidney beans and then cooking the beans themselves to give them a try! It might have been the observation that cooking did make some inedible things edible that drove a hungry family to take the risk, or else the accidental inclusion of kidney beans in a stew that didn't produce horrible stomach pains.

One way or another, though, cooking became part of our everyday process of preparing raw materials to help us produce energy.

The cup that cheers

Of course, eating isn't just about generating energy. Your senses (see page 163) ensure it's an opportunity for pleasure. And there are things you probably consume that have a direct impact on your brain. Take, for example, that essential morning cup of tea or coffee. The caffeine in coffee, tea and some soft drinks is a drug that has a rapid impact on the neural system. Using caffeine to get a mental kick goes back a long way, with tea being consumed for several thousand years in China, while coffee made its way to the West from Africa in the sixteenth century, where the beans had already been used as stimulants for many years.

Caffeine has several effects on the body, but the most interesting is the way that it locks onto receptors in the brain that usually handle a chemical called adenosine. Think of locks and keys – receptors are like locks that will only take certain shaped keys. Caffeine can fit into those intended for adenosine. As it happens, one of adenosine's roles is inducing a feeling of sleepiness, of getting tired. By reducing the amount of adenosine locking onto the adenosine receptors, caffeine can help us feel more awake.

A side effect of the reduced activity of adenosine is an increase in the activity of another natural brain chemical, dopamine. This is a neurotransmitter, one of the molecules used to help carry signals from neurons in the brain to other cells. The result is that familiar little boost that caffeine gives us.

Caffeine turns up in a fair number of plants – tea, cocoa, coffee and cola – providing our familiar stimulating drinks. Its positive attributes for us are an accidental side effect of its intended purpose – caffeine is a natural insecticide that turns up in plants to help kill off predatory insects. It just happens to give us results that we appreciate when it interacts with our nervous system.

It's quite unnerving to think that simply drinking a latte or sipping a cola results in changes in a fundamental operation of your brain. All the evidence is that it causes no damage, though, and has a mildly beneficial effect on the ability to concentrate. Like many drugs it can lead to addiction, and once someone is addicted it will produce withdrawal symptoms if they are denied it. This is why many people who give up coffee claim that they feel better. They are unconsciously comparing the way they feel after the drug is out of their system with the unpleasant withdrawal period. If you like your coffee or tea in reasonable quantities, there's no need to give them up.

Food of the gods

It's sometimes assumed that another favourite product, chocolate, also contains caffeine – but it doesn't. The most notable brain-influencing ingredient of chocolate is a bitter-tasting chemical in the same family as caffeine called theobromine, which in Greek roughly means 'food of the gods'. Theobromine has similar effects to caffeine, though it is somewhat milder and, along with sugars and a melting point similar to the temperature of the mouth, seems to be one of the main reasons we find chocolate so attractive.

It's well known that dogs should not be fed chocolate – this is because theobromine is poisonous to them. A small dog can be killed by as little as 50 grams of strong dark chocolate (which has a much higher theobromine content than does milk chocolate). This isn't a problem that's limited to dogs. It affects all mammals to some extent, though the speed with which the theobromine is disposed of by the system varies from species to species. Cats are particularly sensitive to it, but don't have a problem as they don't have sweet taste receptors, so don't find chocolate particularly appealing.

Theobromine is also poisonous to humans, but this shouldn't cause concern – in a large enough dose practically anything (even water, for example) is poisonous. Humans have about three times the resistance to theobromine per kilogram of bodyweight as dogs, and weigh a lot more, so we won't be damaged by our treats. To get a dangerous dose an adult would have to eat over five kilograms of milk chocolate.

This aspect of dosage being important to toxicity is something you should be aware of if you buy organic food because you are worried about the effects of pesticides on your body. Practically every substance has some risk, but we consume pesticides in such relatively low quantities that their risk is tiny compared with lots of other things we eat. Plants also contain natural pesticides which are just as dangerous to us as the artificial variety.

It certainly makes sense to wash fruit and veg before eating (in part because of bacteria from soil), but if you add up the cancer risk, for example, from a typical diet, 93 per cent of the risk comes from alcohol, and 2.6 per cent from coffee. Once we get the relatively dangerous natural sources of risk like lettuce, pepper, carrots, cinnamon and orange juice out of the way, the main contaminant is a chemical called ETU at 0.05 per cent. If you add up *all* the major chemical contaminants and pesticides at legal levels they have a similar risk to eating celery.

This isn't to say you should avoid eating celery or drinking orange juice – just that it's important to put the risk into proportion.

The winners' drug

Another example of something we take for granted that has significant effects on the brain and body started out as a herbal treatment using the bark of willow trees and extracts of the plant meadowsweet (spiraea). These were used to treat headaches, fevers and inflammations as far back as 2000BC, when they are noted on a medical shopping list dating from the Sumerian Third Dynasty of Ur. They have been popular painkillers ever since.

In the eighteenth century, a misunderstanding resulted in willow bark becoming even more sought after. Peruvian bark or cinchona bark, the source of quinine, was already being used to treat the deadly disease malaria, but it was very expensive. The much cheaper willow bark was recommended as a substitute, but this was a case of confusion. All willow bark does is suppress the symptoms of malaria where cinchona has a more active effect. Still, this error was enough to ensure that willow bark went from strength to strength.

The only problem with the medication was that it played havoc with the stomach. The active ingredient, which we now know as salicylic acid, might have helped headaches and other pains and fevers to go away, but in exchange it disrupted the digestive processes, caused sharp pains in the stomach and could even cause dangerous stomach bleeding.

In 1899, German chemical company Bayer found a partial solution. A product derived from salicylic acid called acetylsalicylic acid had the same medical benefits but was kinder to the stomach. They called it 'aspirin' from a shortened version of the German name for the compound, acetylspirsäure. It became one of Bayer's bestselling brands, alongside their popular cough-suppressant medicine, heroin! And that name 'aspirin' was copyright. Only Bayer could make it. Now, though, it's a generic name in countries like the UK, oddly as a result of a treaty that ended a war.

This was the treaty of Versailles, signed on 28 June 1919. It detailed the reparations that Germany would be expected to pay in ending the First World War. Most of this treaty was the kind of thing you would expect: details of country boundaries, restrictions on military deployment and arms, financial payments and heavy industry provisions. But there, amongst those big concerns, was the right to use the name 'aspirin'.

While in Germany (and 80 other countries around the world) aspirin is still a trademark of the Bayer company, in the UK and other signature countries the name can be used by anyone. It might seem crazy that such a minor right should be written into a major treaty, but both sides had been battered by the terrible Spanish flu pandemic towards the end of the war, and aspirin proved vital in reducing fevers. It was a real medical breakthrough.

Aspirin remained a hugely important drug for over 50 years. When I was a child it was still the only popular over-the-counter painkiller. But by the 1970s it was relegated to an also-ran by the more stomach-friendly paracetamol. This is called acetaminophen in the US, but better known by brand names like Panadol (Bayer's product) and Tylenol. It seemed likely that aspirin would fade from use, that is until it was discovered that it could help prevent heart attacks and strokes.

Aspirin dulls pain and reduces inflammation by disabling an enzyme called cyclooxygenase. Enzymes are special proteins that help chemical reactions in the body go better. In the of case cyclooxygenase, the reaction is the production of a pair of hormones that between them cause inflammation and transmit pain messages to the brain – aspirin works as a painkiller by disrupting this reaction. However, it was discovered that aspirin also reduces the ability of a compound called thromboxane to make platelets in the blood clump together. Platelets are the cells that make blood clot in wounds, but if they clump together in blood vessels they can block the flow, leading to heart attack and stroke. A long-term low dose of aspirin has become commonplace to reduce this risk.

This new use of aspirin has revived the drug's fortunes – around 35,000 tonnes are still consumed each year. Like caffeine this is a relatively simple chemical that interferes with a small part of the complex chemical signalling mechanisms in your body, with beneficial results.

From chemical energy to moving muscle

We might get pleasure or medical benefit from particular chemicals we consume, but the main purpose of eating remains to produce energy. We've already seen how the digestion of food in a slow-burning process generates energy which is stored in ATP molecules. Muscles then make use of this to get you moving. Special proteins release the energy from the ATP, with one protein 'walking' down a filament made of another protein, clawing its way along like a series of little ratchets, resulting in the contraction of the muscle and the ability to move. This process is stimulated by an electrical signal.

An early understanding of this electrical component led to a very famous horror movie. When a young woman called Mary Wollstonecraft Godwin was heading off for a romantic summer holiday with her husband-to-be, she got together a reading list that included a report by the Italian Luigi Galvani on his work. When married, the young woman would become Mary Shelley, and it was on a rainy day during that holiday in Switzerland that she wrote the story that would become the novel *Frankenstein*.

Galvani had been experimenting with dissected frogs, and in what appears to have been an accident, managed to apply an electrical charge to the muscle of a frog's leg, causing the leg to twitch as if it were still alive. Although much misrepresented at the time (not least by Ms Godwin), this was the first glimpse of the importance of electricity within animals as a way of sending control signals.

Making work happen

So far I've just been assuming that 'energy' is a well understood concept, but it's worth clarifying what we are dealing with. We've already seen that energy and matter are interchangeable, though it usually takes a special reaction like nuclear fission, fusion or matter-antimatter annihilation to convert matter into energy. In your body, chemical energy, which is the energy stored in the electron bonds that link atoms together in a molecule, is released to be converted to the mechanical energy of the muscles.

How does having energy make things happen? Energy does work. 'Work' is just energy being transferred from one place to another. When we're shoving matter around, for example, work equates to the force you use times the distance it manages to move something.

Once upon a time, work in the non-scientific sense usually meant toil. Physical labour. These days, what we do in the workplace may be less direct, but even 'brain work' involves a transfer of energy, and often that brain work is a preliminary to physical work. So for instance, in writing a book getting the original idea doesn't involve a lot of physical work, but everything from the typing to the production of a book will. And generally speaking, the body is in the business of converting chemical energy into applied work.

Work, like energy in general, is measured in joules. In everyday life, we often use the old-fashioned energy unit, the calorie, which is a little over four joules. We measure the energy content of food in thousands of calories (or kilocalories). Nutritionists thought that the 'kilo' bit would confuse the public (this was before we were used to metric measurements), so instead of saying something has, say, 129 kilocalories, they often just say it has 129 calories (just plain wrong) or 129 Calories (the capital C is technically correct, but confusing).

The great bumble bee mystery

Every time you move your body, you use up energy from the food you have consumed. That's pretty straightforward. But some creatures give the impression of using more energy in movement than they consume, appearing to generate energy from nowhere. One often-quoted example is the bumble bee. 'It's a mystery,' you will hear. 'No one understands how a bumble bee manages to fly. Science has no answer.' This has even been used from the pulpit as an example of God being capable of something that science can't explain.

In reality, though, the whole bumble bee paradox is a fallacy. Yes, it seems strange that a big fat body can be kept in the air by such tiny fragile wings. But the bee is surprisingly light, and bumble bee wings aren't like a bird's wings. They move at a much higher speed, producing a different lifting effect. The result is more like a set of helicopter blades than flapping wings, and their mechanism produces vortices, fast spinning columns of air that give much more lift than a traditional aerofoil. There simply isn't a problem to solve with the bumble bee. It doesn't make use of more energy than it consumes.

The elastic kangaroo

However, there is one genuine example of an animal that, in a sense, makes use of more energy than it consumes in food. That's the kangaroo. If you add up the energy required to produce all the jumps a kangaroo does in a day, it will genuinely be larger than the energy produced by the food the kangaroo eats. Here is an animal that does produce energy from nowhere – or so it seems.

What the biologists missed when they did the original calculations was that the muscles powering the kangaroo's legs act like a bouncing rubber ball. When you drop a rubber ball and it hits the ground, the ball squashes up, absorbing energy from the collision. It then springs back into shape, releasing that energy and powering the ball back into the air. It's like the energy stored up in a spring or a stretched rubber band. Nothing adds extra energy into the system, but the ball bounces back into the air powered by the energy stored as it is squashed against the floor.

Something similar happens with a kangaroo. The way its muscles are connected up, when it hits the floor, energy is stored up in the muscle, as if a rubber band were being stretched. This is then released and partly powers the next bounce, so it requires much less energy from the kangaroo's food intake to keep it moving. Without this system, all the energy with which the animal hits the floor would be lost in sound and heat. But here some is stored away to use again. It's like the way electric cars push their braking energy into the battery to use it again at a later time.

Heat on the move

Whether we are looking at the energy used in your body, or that used by a kangaroo, we are dealing with thermodynamics. This sounds like the study of heat moving round, and it is, provided you remember that heat is just one form of energy. Heat is the kinetic energy in the moving molecules within a substance – heat something up and those molecules zoom around faster. Thermodynamics became hugely important in the nineteenth century, as it explains how steam engines work and is now a fundamental part of science.

Just how important it is can be seen in a quotation from one of the early twentieth century's greatest scientists, Arthur Eddington. He said 'If someone points out to you that your pet theory of the universe is in disagreement with Maxwell's equations [the equations that describe how electromagnetism works] – then so much the worse for Maxwell's equations. If it is found to be contradicted by observation – well these experimentalists do bungle things sometimes. But if your theory is found to be against the second law of thermodynamics I can give you no hope; there is nothing for it but to collapse in the deepest humiliation.'

We'll come to this 'second law' Eddington was talking about in a moment, but there are a few others. Thermodynamics starts, strangely, with the zeroth law (called this because it was added after the first law, but underpins it). This just says that if you have two objects in contact at the same temperature, there isn't going to be an overall flow of heat from one to the other. As heat is really about bouncing molecules there will be a constant transfer of energy back and forth, but it cancels out.

The first law says that energy is conserved: you can't make it or destroy it. You get out what you put in. The second law, the one Eddington was so concerned about, says that heat (that's to say energy) will move from a hot place to a cold place. And for completeness, there's a third law that says you can't get a body down to absolute zero, the total limit of coldness (see page 36), in a finite number of steps. You can always get a fraction closer, but never make it all the way.

Experiment – Thermodynamics in action

Fill an electric kettle and switch it on. Listen to it and, if it's see-through, watch what goes on. According to the zeroth law, before you switched on the kettle there was no flow of heat between the element and the water. But when you switch the kettle on, the element is heated by electricity and soon is at a very different temperature to the surrounding water. Energy starts to flow from hot to cold, according to the second law.

You should hear a kind hissing noise that gets louder over time. Then, just before the kettle boils, it goes relatively quiet. Right at the end of the cycle you will hear the bubbling noise of the water boiling.

That hissing is the sound of lots of little bubbles of water vapour collapsing. Because the heated element is significantly hotter than the boiling point of water, the water in immediate contact with it absorbs a lot of energy and forms little bubbles of gas. These move into the water, which not far away from the element is much colder, causing the bubbles to collapse with tiny little pops, which add together to make the kettle's distinctive hiss. The sound goes away just before boiling as the water is all practically at boiling point, so the bubbles don't collapse.

Then, as the water reaches boiling point, larger gas bubbles form throughout the water, not only at the point of contact with the element, producing the familiar churning action of a boil.

No perpetual motion machines

The first and second laws of thermodynamics are spoil-sports. Between them they rule out the possibility of creating a perpetual motion machine. If you have young children you might think they *are* perpetual motion machines, but the energy levels of the human body are always being topped up from food. A perpetual motion machine sounds wonderful – there is something magical about the thought of a machine that can go on forever, and you'd only have to connect that machine up to a generator to have an inexhaustible supply of electricity.

If you could break either of those laws, you'd be laughing. If energy conservation doesn't apply, you can use more energy than you put into the system. Once you got a machine started that gave out more energy than you put in, you could start feeding its output back into its input, letting the machine power itself, and still have energy left over. Similarly, if you could break the second law it means you could get energy to go from a cold place to a warmer place. You could then use that energy to do work.

It might seem that a fridge breaks that second law, because it takes energy out of a cold place (inside) and sends it to a warmer place (outside). But a fridge can't do this without help. The second law of thermodynamics only applies if you aren't pumping energy into the system – but that's exactly what you do with a fridge. You use energy to move heat from the cold place to the warmer place, and you use more energy than you move.

People have been trying to build perpetual motion machines for at least 1,300 years. They proved so popular that patent offices stopped accepting patents for them, unless they could be demonstrated with a suitable working model. Sometimes you might see what seems to be a perpetual motion machine, but it will always be getting energy from somewhere outside to keep going.

The energy Crookes

Perhaps the best-known example of apparent perpetual motion is the Crookes radiometer.

This toy has a series of paddles inside a glass bulb. The paddles are not connected to any motive power, and there are no motors or solar cells involved. Yet those paddles go around and around, seemingly forever. It looks like perpetual motion, but really the device is being powered by the Sun – or whatever light source is nearby. The glass bulb may prevent mechanical energy from turning the paddles, but it doesn't stop light, and in this form energy is constantly streaming into the device.

It used to be thought that the radiometer worked by the *impact* of light. One side of each paddle is black, the other side is white. Photons of light will be absorbed by the black side, but will be reflected from the white side. Although photons don't have any mass they do have energy, and as Einstein tells us, mass and energy are interchangeable, so photons *do* have momentum – the oomph to make things move. In fact it would be possible to power a spaceship with huge solar sails, collecting the photon 'wind' from the Sun.

Unfortunately, you need really big sails to get any noticeable push. The paddles in the radiometer are much too small for this. What moves them is the air in the bulb.

This is at relatively low pressure to reduce resistance, but it is there. Because the black side of the paddles absorb photons, they warm up compared with the white sides. Some of this warmth is transferred to nearby air molecules (in accordance with the second law), so the black sides are battered more by faster moving molecules and the paddles start to turn.

Go to the *The Universe Inside You* website **www. universeinsideyou.com**, select *Experiments* and click on *Crookes in action.* The video demonstrates a Crookes radiometer at work.

It's easy to demonstrate that heat is the cause, not light pressure, because the radiometer goes the opposite direction to the one you would otherwise expect. If it moved by light pressure it would push more on the white side, so it would turn with the white sides at the backs of the paddles. But thermodynamics expects the black sides to be pushed and it is indeed these that are at the back of the movement.

Infinite clean energy

Perpetual motion might seem a Victorian idea, but as recently as 2007 a company had a big press launch for what appeared to be a perpetual motion device. Irish company Steorn hit the headlines when it promised to demonstrate a machine that would produce 'infinite clean energy'. The Orbo device, boasted Steorn, used magnetic fields to generate power from nowhere. After much hype, the demonstration of the device in London was cancelled due to 'technical difficulties'. As it was supposedly just a matter of the lighting being too hot for the equipment, causing bearings to fail, it would seem reasonable that a new demonstration would be put on a few days later – but the Orbo has yet to be demonstrated.

Steorn's device supposedly uses a combination of fixed and moving magnets that trace strange paths through the Earth's magnetic field, in order to generate power. Steorn has joined a long line of failures to build a perpetual motion machine – or to put it another way, a truly renewable source of energy. (When we talk about wind or solar power being 'renewable' we really mean they come from the Sun, so will be available for a very, very long time, but they aren't really renewable.)

Entropy increases

The second law of thermodynamics is often phrased in a different way: entropy increases. Entropy sounds like a fuzzy concept when you first hear about it – it's a meas-ure of the disorder in a system. Your body has a lot less entropy than all the chemical components that make you up arranged randomly, because your body has structure. A teenager's bedroom has plenty of entropy. The more disorder, the higher the entropy. In reality, entropy is not just a matter of descriptive handwaving, it's a statistical measure. It's the number of distinguishable ways you can organise the components of the system.

If you think, for instance, of the letters on this page, there is only one way to arrange them that exactly spells out the words you are reading. But there are lots and lots of other ways to arrange them (most of them result-ing in an incomprehensible jumble of letters). So in the arrangement you see the letters are very ordered, with low entropy, and the second law of thermodynamics tells us it took energy to make this happen. They didn't just fall into place, I had to work at it. So did the editors and the printer. There are many random scramblings of these same letters, which would be relatively disordered, with higher entropy.

It seems quite natural that entropy will rise. For example, a teacup is more ordered and has lower entropy than lots of smashed up bits of pottery. It's quite easy to make entropy rise, by dropping the cup and letting it smash to pieces. It's practically impossible to make entropy fall and un-drop the cup, letting all the pieces join back up into a single item.

The idea that entropy rises has been used to argue against the evolution of life of Earth. The Earth has gone from a disordered, random collection of molecules to the comparatively very ordered form that includes all the living things we now see. Some think that this proves there has to be a creator to make order arise from chaos. But that's a misunderstanding of the second law. It says that entropy increases (or stays the same) in a closed system, that it, one where energy can't get in or out. The Earth isn't a closed system. We have a huge amount of energy pouring in from the Sun, and that's how we can flout the second law.

The physics of monsters

This link between entropy and life isn't the only example of basic physics having a direct effect on living things. Your body has no problem with the laws of physics (apart, perhaps from a little sagging duo to gravity).

But the same doesn't apply to the typical monster. A traditional scary monster of fantasy fiction is the oversized insect or spider. A ruthless killer like a spider would be horrific if blown up to a size where a human could be its prey. Thinking about it, it might seem odd that we aren't faced with such scary predators – why haven't they evolved to take over the world? Think of the massive ants of the 1950s movie *Them!* or the huge spiders in *Lord of the Rings* and *Harry Potter*.

If such creatures give you nightmares, you can reassure yourself with the knowledge that they can never exist. Imagine we took the sort of spider you find in the bath and made it 100 times bigger. Horrible! What we mean by '100 times bigger' is that it would be 100 times as wide, with legs 100 times longer. If we cut through such a spider's legs, in cross-section they would be 100 × 100 – 10,000 times the area of a normal spider's legs.

But what about the weight of such a titanic arachnid? Weight depends on volume, so the spider would be 100 × 100 × 100 – a million times heavier. This means you end up with a million times the weight being supported by legs that are just 10,000 times greater in cross-section – the spider would collapse under its own weight.

This would also happen to giant mammals (including human beings), but spiders and insects would have another problem if blown up to enormous size. That's because they breathe through their carapace (their hard outer 'skin'). The surface area of the carapace, and so the amount of oxygen taken in, would go up by 10,000 times but would be supporting 1,000,000 times as much body. Such a massive spider would therefore die of asphyxiation at the same time as all its legs snapped. Not quite so scary.

Staying on two legs

Back at your body, your means of locomotion is, you might think, simpler than a spider's. After all, a spider has eight legs to coordinate in order to avoid tripping over its own feet. It's certainly true that with two legs we have relatively few gaits – different leg movements – to learn. But we have a different problem.

Being perched on two legs is not very stable; you only have to watch a toddler taking its first steps to see this. It's a bit like the difference between riding a bicycle and a tricycle. When first learning, the bicycle rider has the additional problem of balance, and so does anyone attempting to walk on two legs. This takes a fair amount of practice and also consumes energy. In practice, walking on two legs is a kind of repeated fall combined with recovery.

Fidgets and knuckle-crackers

Even standing up is a drain on your energy – significantly more so than sitting down – and few of us can really sit still. It's fascinating to watch other people when they are supposedly still, and seeing just how much they fidget, shift and generally make small movements. Hands are particularly good at filling in time by staying on the move. My grandmother often twiddled her thumbs – an easy habit to fall into, though it's hard to explain why. And then there are the knuckle-crackers.

The standard defence for those of us who don't crack our knuckles is to warn offenders that they will end up with arthritis if they keep doing it. But is it true? One man has made it his mission to find out. Donald L. Unger, a medical doctor from Thousand Oaks, California, has been cracking his left knuckles every day for over 60 years while leaving his right hand untouched.

Although it's impossible to draw definitive conclusions from a single person – Dr Unger could be like one of those people who regularly crop up on the news saying 'I have smoked 40 cigarettes a day since I was 20, and I'm now 95' – he has not suffered from any disadvantage to his left hand, so it could be that the linkage of knuckle cracking and arthritis is an old wives' tale.

Standing around, with or without knuckle-cracking, is certainly a capability that will be tested in full at our next destination. I'd like you to head off to a theme park.

6. Feeling dizzy

Take a ride on a modern rollercoaster, the kind where you are flipped upside down, spun around in a corkscrew and generally flung about and disoriented. What happens to your view of the world? How does it impact your senses? When you get off you may well be feeling dizzy and a little wobbly on your feet. So, why does taking a ride continue to mangle your senses, even after you get off it?

Senses are one of those biological processes that define life. They provide the interface between you and the world. Without your senses you have no way of finding anything out, no way of reacting to the world around you.

Counting the senses

How many senses do you have? The knee-jerk response is five, but this is so obviously wrong when you think of that theme park ride. Which of the traditional five senses – sight, hearing, smell, touch and taste – was the one that told you that you were upside down? You might think it was sight, and that certainly contributes, but do you really think you wouldn't notice that you were upside down if your eyes were closed? The big five are important, but they are just the beginning of our exploration of the senses.

We've already looked at sight in action when gazing at the stars, but what is hearing all about?

Sound is often described as a wave, but it is more accurate to call it a regular sequence of pulses. Our typical picture of a wave is something like a ripple on the surface of water. The waving takes place side to side, at right angles to the direction that the ripple moves along. But in a sound wave, the waving is in the same direction that the sound is moving.

Experiment – Simulating sound

Get hold of a Slinky spring toy and fix one end to something that will hold it in place (alternatively, get someone to hold the other end). Stretch the spring out horizontally until it is reasonably taut. Now give the end you are holding a sudden sharp push forward, followed immediately by a tug back.

You should send a ripple down the spring towards the other end. Try this several times. As the wave heads down the spring, the spiral of metal compresses, then stretches out.

If you don't have a slinky, visit **www.universe insideyou.com**, click on *Experiments* and then choose the *Waves in Springs* experiment to see a sound-type longitudinal wave in action.

This kind of 'longitudinal' compression wave is what sound is all about. When something vibrates – a loudspeaker, a musical instrument or your vocal cords, for example – it pushes on the nearest air molecules. They are squeezed closer to the next lot of air molecules, just like the spring compressing. This next layer of air molecules are repelled and move away, getting closer to the next lot of air molecules … and so on. The compression travels forward. This is sound, travelling at around 340 metres per second through the air.

When you think about it, sound has to be this sort of compressing wave, rather than a side-to-side wave like a ripple on the water. If it went side to side as it travelled, then it would keep bumping into the rest of the air and would soon run out of energy. Usually side-to-side (or 'transverse') waves can only run along the edge of the stuff that's doing the waving. The only exception is light – if you think of this as a wave, it does go side to side and can travel through the middle of a medium, but that's because it isn't a wave in *stuff* – it works just as well in the vacuum of space.

The speed of sound is one of the basics of nature that it is easy to test out yourself. Next time there's a thunderstorm you can measure the speed of sound directly. Thunder is simply the sound made by the air as a flash of lightning sends temperatures soaring to as high as 20,000°C. Both thunder and lightning occur simultaneously at the same place.

If you see a flash of lightning, then hear the thunder some time after, you can use this to get a direct feel for the speed of sound. To all intents and purposes the light will arrive instantly. If the thunderstorm is, say, ten kilometres away, the light takes just 1/300,000th of a second to cover the distance. So the delay in hearing the thunder tells you how long the sound takes to cross the distance. For that ten kilometre distant storm, travelling at around 340 metres per second, the sound will lag behind the light by over 29 seconds.

From compression wave to brain wave

Eventually, after pulsing its way through the air, a sound will arrive at your ear. The outer, visible part of the ear funnels a wide spread of the compressions into the small hole in the side of your head, amplifying the wave. Further into your head, the compression hits your eardrum. This membrane moves back and forth as the shifting air molecules push and pull it. The eardrum passes the movement through three tiny bones – the smallest in the body – onto a second membrane called the oval window, which sets fluid in the cochlea in motion.

The cochlea is a spiral-shaped bone chamber (the word comes from the Latin for a snail), which is filled with a watery fluid. The movement in the fluid is picked up by tiny tufts which look like hairs, but which are actually extensions of cell membranes. These stimulate the 'hair cells' at their base and generate signals in the auditory nerve. Just like in sight, what had been an external physical phenomenon has became an electrical signal travelling through a nerve to the brain, where it will be processed to build up a sound picture.

Some people have damage to these hair cells, and their hearing can be partially restored with a cochlear implant. This directly stimulates the neurons that the hair cells should act on. An external headset picks up sounds and processes the signal to produce a series of electrical impulses which are transmitted to a small device implanted under the skin, which in turn stimulates electrodes embedded in the cochlea. The earliest implants only had a single electrode, though numbers have increased over time to twenty plus separate stimulation points. Even though this means that only a small subset of the hair cell-connected nerves are activated, the implants have enabled users to understand speech, proving much more effective than was originally expected. Over 100,000 people have now benefited from cochlear implants.

Audible illusions

We tend to think of hearing as a more straightforward sense than sight. There are so many well-known optical illusions that it's not difficult to accept that the brain constructs a visual chimera that is its attempt at building an image from the various inputs it receives. Sound, though, we tend to think as just, well … sound. We assume that what we hear is what's out there. But again the incoming raw data is subject to processing and manipulation by your brain.

It is entirely possible to produce an auditory illusion. If you would like to experience one for yourself, go to **www.universeinsideyou.com**, click on *Experiments* and select the *McGurk Effect*. Follow the instructions on the web page.

The sound of emotion

Like sight, sound is much more than a simple source of information; it can powerfully influence our emotions. If there is a moment in a film that brings you close to tears it is liable to be the sudden swell in the music that triggers the emotional response. Even the absence of music can be effective in this way. If a drama makes heavy use of a soundtrack, a sudden period of silence can build tension and add a feeling of real involvement.

Another example of the effect sound has on our emotions is when we find a noise irritating. An irritating sound can dominate our senses. The most famous (and universally cited) irritating noise is the sound produced by dragging your fingernails over a blackboard or slate. Analysis has been done on the impact of this sound, and unexpectedly it is not the high frequencies that make the sound so distinctive that upset us. These can be taken out and the result is still grating on the ear.

It has been suggested that the sound of fingernails on a blackboard might be similar to a pre-human warning cry (the frequency distribution of the disturbing part of the noise is similar to that in the warning cry of macaque monkeys), or the call of some long-forgotten predator. Whatever the cause, as the research paper on it concludes 'the human brain obviously still registers a strong vestigial response to this chilling sound'.

All in good taste

Sight and hearing are the headline senses. They are the ones that have a major impact on your life if you lose them. But taste is rather different. Okay, it helps us to tell if we've eaten something unpleasant, and it turns eating from a chore into a pleasure. But it's not an earth-shattering sense. What's more, taste is also by far the least effective of the five senses. Its first problem is simple access – before you can taste something it has to get into your mouth – which limits its scope. But taste is also more limited because you rely for a considerable extent on your other senses to supplement your sense of taste.

Experiment – The limitations of taste

This experiment requires a little preparation. Chill two glasses of wine (adults only, this one, I'm afraid), one red and one white, to the same temperature in the fridge. While the wine chills, cut up some small pieces of food with similar textures but very different tastes. Try, for example, a few different cheeses, some raw fruit and vegetables, chocolate and a heavy bread.

Now block your nose using screwed up tissue paper (assuming you don't have nose plugs) and cover your eyes with a mask. You may need help to carry out the experiment at this point. Ideally have someone else mix up the samples before you try them, as you can easily fool yourself if you know which is which.

Take a sip of each of the glasses of wine. Most of us think we can tell the difference between red and white wine, but is it as obvious without sight and smell?

Try the different food samples. They *will* taste different, but are they as distinctive as usual without your senses of smell and sight making a contribution?

A lot that you habitually consider to be taste is in fact smell, or is influenced by what you can see. When the taste buds are left to their own devices, they don't manage anywhere near as well as expected.

There also seem to be some differences in how we perceive taste that are influenced by sound. Loud background noise seems to make us think that the food we are eating is less sweet and salty, but more crunchy.

When eating crisps (potato chips), when the sound of a loud, crunchy bite was played to them as they ate, participants in a test thought that the crisps they were eating were fresher and crisper than those who heard a quieter sound.

Flavours and taste buds

There are five key flavours detected by the taste buds – sweet, bitter, sour, salty, and umami. This last one is the least familiar. It is the 'savoury' flavour, often described as the one that appears in a concentrated form in the flavour enhancer monosodium glutamate, often added to processed foods to bring out the savoury taste. You may well have seen one of these diagrams, mapping out which areas of your tongue detect the different flavours.

I have even seen experiments where it has been claimed that subjects can detect those flavours when the particular area of the tongue is stimulated by pressure.

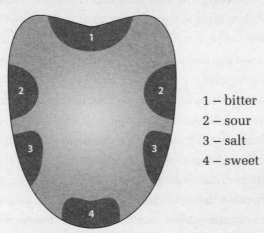

1 – bitter
2 – sour
3 – salt
4 – sweet

Tongue flavour map: now discredited

This appears to be a totally fictional idea, dating back to the early part of the nineteenth century. It is no more real than phrenology, the Victorian idea that your mental capabilities are reflected by the size of the bumps on your skull. The reality is that every part of the tongue can detect all the flavours.

There are somewhere between 2,000 and 6,000 taste buds on the tongue. Each is a small depression in the surface, through which foodstuffs (dissolved in saliva, if solid) can come into contact with taste receptors, which produce signals as a result of the presence of certain chemicals. So saltiness, for example, is primarily the detection of sodium ions, while sourness is the result of the receptors detecting an acid.

The mineral in the kitchen cupboard

Salt is a most unusual part of our diet. If you search your food shelves, salt is likely to be the only item that you regularly consume that has not come from a living thing – it's a mineral. It is also a very rich component of our vocabulary. Are you worth your salt, or the salt of the earth? Are you nasty enough to rub salt into someone's wounds, or rich enough to salt away a fortune?

Salt, the simple compound sodium chloride, combines two dramatic elements. Sodium is a metal that practically explodes on contact with water, while chlorine is a green poisonous gas, the first widely used chemical warfare agent, that wreaked havoc in the First World War. That the two should come together to make those stable little white crystals is quite surprising.

All animals need salt in small quantities (though it's not totally obvious why it is one of the big five tastes). It acts as an electrolyte – a carrier of electricity in a fluid – in the body, which means that it has always featured as part of our diet, though the chances are that to begin with all the salt human beings consumed was mixed up with other things, for example in animal blood.

As an aside, you'll sometimes hear that Roman soldiers were paid in salt, and this is where we get our word 'salary' from. It's certainly the source of the word (salt is *sal* in Latin), but the soldiers' *salarium* was a payment in ordinary money for the purpose of buying salt. They didn't receive a wage packet containing a chunk of rock salt, however attractive the picture may be.

Salt has a very distinctive taste, and there are few things we experience that have more of a salty kick than an accidental mouthful of seawater. But strangely seawater does not contain salt! Seawater does contain, in solution, both sodium ions (see page 34 for more on ions) that are from dissolved rock material and chloride ions, mainly originating from underwater volcanoes and vents. But the two sets of ions drift around independently. It's only when seawater is evaporated that sodium chloride – salt – forms. In principle, water would taste salty if it just had sodium ions in it (or indeed ions of the similar metal potassium).

Sniffing your way around

Like taste, your sense of smell has less day-to-day value than some of your other senses. Yes, it's helpful to detect smoke, or to find something around the house that has gone off. And as we've seen, it is a major contributor to the pleasures of taste. Yet smell remains a limited sense, for humans, at least. Although it can detect something at a distance, it's very difficult to get any feel for direction with smell – and all too often this sense is crippled by colds and other infections that block the nose.

Your sense of smell is quite closely related to taste, not only in the way that you use it, but also in the way it works. Smell is a process of detecting various chemicals, using specialised detectors at the back of your nose and above it, inside your head. Chemicals carried on the air dissolve in the mucus above the receptors and the captured chemicals interact with those receptors, triggering signals to your brain.

In the animal kingdom, smell can be much more important than it is to us. You only have to watch a dog taking a walk in the park to realise this. Certainly a dog uses its eyes and ears, but its nose – vastly more sensitive than ours, capable of picking up smells that are well over a million times more dilute than anything we can sniff out – is also constantly in action. To a dog, the scentscape of the park is just as important as anything it can see.

Scenting a mate

Smell isn't just used to detect threats and prey, it can also be a way of communicating with other members of the same species. Dogs spend more time when out and about picking up the scent from other dogs than doing practically anything else. The best-known (though not the only) chemicals in the smelly signalling business are hormones called pheromones. Insects that act together as if they were a larger organism, such as bees and termites, have a wide toolkit of pheromones for signalling and coordinating actions.

Humans produce pheromones too, though there is much debate about how much smell influences our attitude to the opposite sex. One famous experiment looked at the way differences in a particular gene might change preference via the production of hormones and the sense of smell. It had been discovered that various animals tend to sniff out a mate that has different variants of a particular group of genes, HLA, that are partly responsible for the ability to resist infection.

The implication was that it was beneficial for offspring to have different variants of HLA, giving them a better chance of fighting off bugs. If animals did this, could it also influence human choice? The experiment, undertaken in 1995, asked a group of women to sniff T-shirts, each of which had been worn for two nights by a different man. After testing the genes of those involved it turned out that the women had a preference, based on smell alone, for different HLA genes from their own.

So it's possible that our sense of smell does influence our selection of a partner – though it should be stressed that this isn't the only factor involved. For example, we also show distinct preferences based on face shape. And in opposition to our sense of smell, we seem to select face shapes of potential partners with HLA genes that are *close* to our own. We certainly aren't at the mercy of a single genetic impulse through scent, but it does seem to have an influence on human attraction.

À la recherche de odeur perdu

You will often hear it said that smell is is stronger than any other sense when it comes to evoking detailed memories. This turns out to be a myth – there is no evidence that smell is better at triggering memories than any other sense. However, it does seem from the way neurons fire in the brain that the first time a smell gets tied to a particular object or event it kicks off much more energetic brain activity than on subsequent occasions, which means we may well remember the first time we experienced a particular smell.

This isn't the case with other senses, and it may mean we have a greater tendency to associate smells with early (and hence evocative) memories than we do any other senses. So when Marcel Proust tediously droned on about childhood memories evoked by the taste of a madeleine cake dipped in tea in *À la recherche du temps perdu*, he used the wrong sense as a trigger.

The sense that's everywhere

Smell, like the other three senses we have so far covered, is concentrated on a collection of receptors in a small area of the body. But the fifth sense, touch, is much more diffuse. Although you have a more developed sense of touch in some parts of your body than others, all of your skin is equipped with touch receptors.

Touch is primarily a mechanical process. Sensors in your skin react to pressure or to deformation of the skin's surface. Touch is distinct from the other four main senses in that it isn't a means of detecting an incoming trigger – light, sound or chemicals – but instead it monitors changes in the body itself. The other senses focus on the environment, but touch keeps tabs on your body.

Seeing with your skin

So with five senses in the bag, why do we need yet more? Here's a simple example: put your hand a few centimetres away from an iron that is switched on. There is nothing that your five senses can tell you by, say, looking at the base of that iron, to let you know that it will burn you. Yet you can feel that the iron is hot from a distance, and won't touch it if you are sensible. (That's a particular good word here. 'Sensible' originally meant detectable by the senses.) How do you pick up the heat from a distance? Because your skin has sensors that detect an invisible form of light, infrared.

Surprisingly there doesn't isn't too much known about exactly how you feel temperature, though it is thought that there are different mechanisms for detecting hot and cold things, and there may be separate mechanisms both to deal with overall temperature (does an object feel hot or cold?) and the direct impact of infrared on the skin. Clearly there is some kind of receptor in the skin which enables us to judge the impact of heat as you experience it when near anything hot, but the details are yet to be established.

A sense of pain

We speak of spicy food like chilli or curry as being 'hot'. Yet this experience doesn't use the same sense that detects the heat from an iron. Nor is it taste. The *taste* of a chilli pepper is not dissimilar to a bell pepper – those mild red, yellow, orange or green peppers you find sliced in salads. But when you bite into a chilli pepper the taste is wiped out by a new sense coming into play – the sense of pain.

Although many of us enjoy eating spicy food, the sensation that such 'heat' triggers is really pain. Chilli peppers contain capsaicin and other substances which bind onto pain receptors in the mouth. They have similar effects on delicate areas of skin like the eyes, should they come into contact, hence pepper spray, which contains capsaicin. Incidentally, pepper spray would be no use if you were being attacked by ostriches – birds don't have a receptor for capsaicin so it has no effect on them.

A chilli is just one of many ways to produce pain, and a relatively mild one at that. Pain is a complex term that really covers several different senses. You will experience pain as a result of chemical sensors, like those that pick up the capsaicin, but also through heat and mechanical sensors when the amount of stimulus they receive goes over a trigger level. A small amount of heat is pleasant, but over the trigger point you are burning – it become pain.

Similarly, a mild mechanical stimulus to your body is just felt as a touch. But if it goes too far – say something sticking into you and distorting the flesh – it becomes pain. Like all the senses, a stimulus at the original receptor generates a signal in the nervous system – the internet of the body. This signal is routed to the brain, and it is only there that what was a simply a chemical and then electrical signal becomes a sensory experience of pain.

That's how painkiller tablets work – not by travelling through the body to the point where you experience pain and somehow interacting with the part that is damaged, but by intercepting the action in the brain and stopping the pain signal getting through.

Pain has an important function, but the way it works is an example of the 'design' of the body being not quite right. Pain could be a whole lot better than it is. We need to be alerted to sources of pain, but the sense is often out of proportion to the urgency of the problem. Although we can block some other senses (think blindfolds and ear-plugs), pain is the sense we most often try to modify for our personal benefit. If human beings were truly designed rather than evolved, a good designer would give us an easy way to switch off pain once it has done its job.

Finding your own nose

Experiment – Participate in proprioception

Here's a very simple experiment that demonstrates another of the ways that you have sensory capabilities that go beyond the famous five. Sit down and close your eyes. Stay still for a moment with your hands by your sides. Now bring up one of your hands and, using your forefinger, touch the tip of your nose. Try it before reading on.

Unless they have brain damage, most people can do this easily. Clearly you need a sense to be able to do this, but which of the five senses helped you? None of them. This is a whole different mechanism.

The experiment above engages the most obscure and indirect of your senses, called proprioception. This is the sense that detects where the parts of your body are with respect to each other. It's a kind of meta-sense, combining your brain's knowledge of what your muscles are doing with a feel for the size and shape of your body. As the experiment you just did shows, this is a mechanism that enables you, without using your basic five senses, to guide a hand to unerringly touch another part of your body.

Other animals have an even wider range of senses than we do. Sharks, for instance, can detect the electric fields generated by the nervous systems of prey, while some birds detect the Earth's magnetic field to guide their migrations – in effect they have a built-in compass. Animals that use echolocation, like bats, may use the same sensors as we do for hearing, but they are employing a totally different sense, one that constructs something closer to the three-dimensional experience of vision than simply hearing noises.

Sensing the accelerator

The sense that particularly comes into play as you are hurtled around on the roller coaster that opened this chapter is one reliant on acceleration. It's often identified as part of the sense of balance (given the fancy name equilibrioreception), but this is an example of biologists confusing sense and function. The main *use* of our acceleration detection is to help with balance, but what we *sense* is acceleration.

You can do this because your inner ear has fluid inside that sloshes around with movement. This flows over little hair cells that are pulled with the movement of the fluid and signal to your brain how you are moving. This is the body's equivalent of the accelerometers in modern mobile phones that enable them to tell how they are turned and twisted. It's also the reason that you are left dizzy and shaken when you get off the rollercoaster – fluid is an effective acceleration sensing mechanism for a biological system, but the downside is that it takes a while to stabilise after being seriously disrupted.

On the rollercoaster you are subject to two main forces – gravity, pulling you towards the centre of the Earth, and the force the car imposes on you, pushing and pulling you in all directions as you hurtle around the track. This pushing and pulling is often referred to as g-force ('g' for gravity).

Weight and mass

In effect, the g-force you feel is a kind of artificial weight. Weight is a word we have to be careful with in science. The weight of something is the amount of force it feels due to gravity, but we tend to use it as an alternative word for mass, which is a measure of the amount of stuff there is in something.

It's easy to get confused, because we use the same units for weight and mass – but they are fundamentally different. A one kilogram weight bag of coffee on the Moon would contain six times as much coffee as a one kilogram weight bag on the Earth. But a kilogram mass of coffee would be the same.

You may weigh 70 kilograms on the surface of the Earth, but up on the International Space Station, your weight would be practically zero. Your mass would also be 70 kilograms on Earth, but this value would stay the same up on the space station. This shouldn't be a surprise; as we have already said, mass simply describes how much stuff there is in you. That stuff doesn't disappear just because you go into orbit.

As well as saying how much stuff there is in something, mass tells us how much force it takes to get an object moving, something that Newton worked out in his second law of motion (in fact Newton invented the concept of mass for this purpose). The second law says that the amount of force you need to get something moving is the object's mass times the acceleration it experiences. So the faster you accelerate something the more force it takes to do so. And this is exactly the same whether you are on the Earth or in space.

Weight is the amount of force gravity applies to a quantity of mass. The acceleration caused by gravity on the Earth's surface is around 9.8 metres per second, every second. So if you fall, every second you will get 9.8 metres per second faster. The force – your weight – is just 9.8 times your mass – but we fiddle the units and measure weight in the same units as mass. Weight should really be measured in newtons, the unit of force in the metric system – when we ask how much a new baby weighs, the answer should really be something like 35 newtons, but this would confuse everybody.

When you experience g-force on a rollercoaster (or anything else that's accelerating) it might be, say, 2g – two times the force of gravity. This isn't a scientific unit, but it's helpful to give an idea of how it feels. Without special equipment to provide support, forces up to around 5g are tolerable by humans, and over very short periods of time people have survived sudden shocks of up to 100g.

Push me pull you

If you imagine going around a corner on a theme park ride (or in a car), it's not obvious which way the force is acting on you. If the vehicle turns right you feel that you are being pushed against the left side of the car. You seem to be flung outwards by 'centrifugal force' – but this feeling is misleading. There is no such thing as centrifugal force. Common sense says, 'Yes there is, that's how I ended up sitting in my neighbour's lap after that tight turn,' but physics knows better.

It was Newton who spotted what was really happening. Once something is moving it keeps going in a straight line, unless you push it to change its direction, or push it to slow it down. It just so happens that everyday objects on the Earth are always being given a push to change direction by gravity (for instance when you throw a ball, it goes from travelling horizontally to curving down towards the ground), and are always given a push to stop, thanks to friction. (Things can also be given a push to change direction by spinning them, as when a football is 'bent' around a wall of players.)

So now let's get back to that imaginary centrifugal force. Let's say you've moved from the rollercoaster onto one of those teacup rides at the theme park. As you are spun round, it feels like there's something pushing you outwards. But once you start moving, it takes no force to make your body carry on in a straight line (though in practice a force is required to counteract friction). You don't need a force to push you outwards.

Instead, the outside of the teacup stops you from heading outwards and pushes inwards on you to keep you in the cup. The force is not actually outwards (centrifugal) but inwards (centripetal – a term Newton invented), resisting your natural tendency to travel in a straight line. It's the same on the rollercoaster or in the car. Once your body gets moving it will try to head off out of the car in a straight line, but the car pushes against you in the direction of the turn, applying force to change your direction.

The occult force

As you sit reading this book, the gravitational force is the most obvious one that you feel acting directly on your body. This pull towards the Earth is one of the four forces of nature (we'll meet the other three in a few pages), and it acts at a distance, something which worried scientists for hundreds of years. When Newton described the way gravity kept the planets in their orbits in his (frankly almost unreadable) masterpiece *Philosophiae Naturalis Principia Mathematica*, he was mocked by a lot of his contemporaries because he described gravity as an attraction, just as one person is attracted to another. His ideas were called 'occult' and 'absurd'.

The problem was that, generally speaking, to make something happen at a distance you have to send something from A to B. If you want to make a can fall off a fence, you have to throw something at it – you can't just think at it and make it move. If you want me to hear you,

you have to send sound waves through the air between us. But gravity seems to work without anything connecting the bodies that are attracting each other. Newton just shrugged his shoulders and said 'I frame no hypotheses.' He had no idea how gravity worked, but he knew his maths did the trick and tied together everything, from falling apples to orbiting planets.

Warping space and time

The man who took away the problem of how gravity manages to keep your body in place on the Earth was Albert Einstein. If you ask people in the street what he's most famous for they'd probably say $E=mc^2$, his equation linking mass and energy, which is certainly important. But ask any scientist and they will tell you his most impressive bit of work was 'general relativity', his theory of how gravity works.

It's infamously complicated, at least as far as the maths is concerned. Even Einstein had trouble with it and had to get help from better mathematicians. But the basic principle behind it is so simple it almost seems trivial. Einstein dreamed it up in his spare time at work in 1907. Here's how he described that moment: 'I was sitting in a chair in the patent office at Bern when all of a sudden a thought occurred to me: "If a person falls freely he will not feel his own weight." I was startled. The simple thought made a deep impression on me. It impelled me toward a theory of gravitation.' He would later describe this as 'the happiest thought of my life'.

Einstein had come up with the principle of equivalence. It says that gravity and being accelerated are identical. If you are inside a rocket, say, with no windows, and feel yourself pulled down towards the floor, there is no experiment you can do that will tell you whether that pull is caused by gravity or because you are experiencing the g-force of the ship accelerating. There is no way to tell them apart.

Of course you can cheat. You can use something like GPS to pinpoint your location and acceleration. Or you could do experiments in different parts of the ship. If the pull you feel is gravity, it should vary between a part of the ship that's further from the Earth and one that's nearer. But that's not what Einstein meant. If you make a measurement at a particular point, and don't use technology that can look outside the ship, there is no way of telling if the force is caused by acceleration or gravity.

Because the two are equivalent, you can use an acceleration to counter gravity – in effect to switch it off. This is what happens in free fall. You have probably seen footage of 'vomit comets', the aircraft that climb high into the air then accelerate towards the ground at just the right speed to counter gravity, leaving the occupants floating for maybe twenty seconds before the plane has to pull out of its dive.

Falling and missing

It is also, less obviously, what would happen to you if you visited the International Space Station and observed the effect on your body. Astronauts experience practically no gravity, but this isn't because they are far away from Earth. At the height the ISS orbits, gravity is around 90 per cent of its ground-level strength. But by the nature of its orbit the station and its occupants are constantly falling at the right speed to cancel out that gravity. The only reason they aren't burned up or smashed to pieces is that they keep missing the Earth.

In its orbit, the station flies sideways as well as falling. The two movements cancel out, keeping it at the same height, but still in free fall. Once he started thinking about this equivalence, Einstein had another wonderful

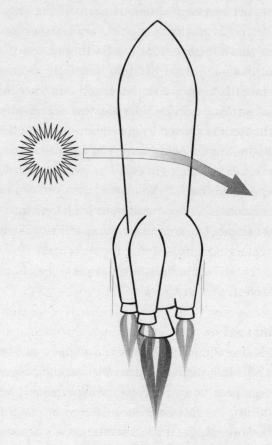

Light crossing a spaceship

thought – if you shoot a beam of light sideways in a space-ship that's accelerating, the beam will be left behind. In effect it will bend as it crosses the ship. But if gravity is indistinguishable from acceleration, it too should also make light bend.

A lesser mind than Einstein's might have decided that light would be pulled by gravity in the same way that other things are. But instead this inspired a totally wild idea – what if a massive object like the Earth didn't attract things, what if it twisted both space and time (space-time)? This would result in the light beam bending.

The image that's often used of general relativity's action is a bowling ball on a rubber sheet. That sheet represents space and time. The ball causes a depression in the sheet. If you imagine a beam of light as a straight line running through the sheet when the ball is put in place, the line will bend – the light will now travel around a curve. The mass has warped space and time and changed the light's direction. As far as the light is concerned, it is still going along in a straight line, though; it's the space-time the light moves through that is curved.

No more action at a distance

The neat thing about this general relativity approach to gravity is that it does away with the messy need for action at a distance. Anything with mass warps the space-time around it, and this distortion spreads through the fabric of space-time. Even your body creates its own, very small, distortion in space and time. So when anything encounters a part of space-time that has been warped it will feel the pull of gravity.

The rubber sheet model is fine, but can be a bit misleading. Firstly, its picture of space-time is two-dimensional, but the real thing has three dimensions of space and one of time. And the rubber sheet isn't so good for explaining why an object starts moving (Newton's apple, for instance) when it feels the gravitational pull of something else.

This has been described as being like putting a ball bearing on the edge of the dip in the rubber sheet caused by the bowling ball. The ball bearing will start rolling down the slope towards the ball. But why does the ball start rolling? What makes it roll down? Erm, well, it's gravity. So this explanation uses gravity to explain gravity – it's a circular argument and is useless.

The reality is much more startling. Take an apple, hold it at waist height and let go. The moment you let go, the apple feels the pull of gravity. The apple is pulled towards the Earth and the Earth towards the apple. But the Earth has much more mass, and a much bigger pull. Soon the apple is dropping, getting faster and faster.

How did that happen according to general relativity? It's all in the fact that the mass of the Earth doesn't just warp space, it warps *space-time*. Although the apple was stationary in space, it was moving through time. Once the space-time was warped, that movement in time had to be partly twisted into another dimension. But there is only one dimension of time – so some of the movement through time became a movement through space. The apple is accelerated through space and falls to the ground because of a warp in time. Mind-boggling, but true.

Slowing your clocks

You might think this means that you lose some of the movement through time, so time should run slower in a gravitational field – and it does. GPS satellites work by comparing times on accurate clocks. These times have to be corrected to deal with the shift caused by Einstein's other great theory, special relativity. This says that time goes slower on a moving object – so the clocks on the satellite are a little slower than those on Earth. But the satellites also experience less gravitational pull than they would if they were on Earth's surface. So general relativity says their clocks will run fast, and this is the biggest correction that has to be made to make GPS work.

Experiment – Evaluating equivalence

Get hold of a helium-filled balloon on a string and take it for a ride in a car. (Get someone else to drive.) Hold the balloon string roughly central in the car's passenger compartment, with the balloon floating, but not touching the car's ceiling. When it's safe to do so, ask the driver to brake. This should be steady braking for several seconds rather than a sudden slamming on of the brakes. What happens to the balloon?

A quick assessment of what's happening in the experiment above might suggest that the balloon should head towards the windscreen. The argument is something like this: when the driver applies the brakes, the car slows down, or to put it another way, it accelerates in the opposite

direction to the way it's going. (Deceleration is just acceleration in the opposite direction to the movement.) That
acceleration isn't applied to the balloon, so the balloon
continues moving forward. This is Newton's first law of
motion, which says that unless you apply a force to it,
something will continue moving the way it was.

In fact, something quite different will happen. The easiest way of understanding it is to make use of Einstein's
principle of equivalence. As it slows down, the car is
accelerating towards the back. Because acceleration and
gravity are equivalent, this is the equivalent of there being
a gravitational pull towards the front of the car. (Think
of the original rocket example. You feel the same pull if
there is gravity *downwards* or if the rocket is accelerating
upwards.) When the car brakes you are pulled towards
the front by this 'gravity' caused by acceleration.

Now think what a helium balloon does when it experiences a normal gravitational pull towards the ground.
It goes in the opposite direction to the gravity, because
it weighs less than the air it pushes out of the way, so it
feels a force in the opposite direction to gravity (usually
called uplift). This means that if there is a gravitational
pull towards the front of the car, the helium balloon will
float off towards the back of the car – and this is indeed
what it does.

The force of creation

Gravity is the most obvious of the four forces that make it
possible for your body to exist. You can't miss it. Without
gravity there are so many reasons you wouldn't be here.
It's not just a matter of keeping you on the surface of the

planet, or keeping the Earth on its path around the Sun. It was gravity that formed the Sun and planets in the first place. Around 4.5 billion years ago, what had been a cloud of dust and gas had been pulled together sufficiently for the Sun and planets to form, all under the influence of gravity. This ubiquitous force also set the Sun in action, producing the heat and pressure that are essential components of the nuclear fusion that produces all our heat and light.

There are more subtle benefits we get from gravity too. Astronauts who spend too long in space find that their muscles are wasting away and that their bones are becoming weaker and weaker. It's quite possible that we couldn't live our whole lives without gravity. Apart from anything else, it's harder to breathe without that steady downward force – your liver floats up, squashing the lungs, and the diaphragm shifts, cutting down on lung capacity. A baby born in space might not survive because of this.

Certainly other living things struggle in zero g. It has been known for a long time (Darwin noticed it) that plants depend on gravity to know where to send their roots.

In space, roots lose all sense of direction and straggle all over the place. Birds' eggs have even bigger problems. In an experiment on the International Space Station (bizarrely sponsored by KFC) it was discovered that yolks that aren't held near the shell by gravity don't develop properly, so the birds don't hatch out.

The force of electricity and magnetism

Yet despite being in-your-face and important to the universe and to human life, gravity is by far the weakest of the four forces. This becomes obvious when you compare it with the other 'everyday' force that has a very obvious impact on your body, electromagnetism. As the name suggests, the electromagnetic force is responsible for electricity and magnetism. But that doesn't limit it to hairdryers and fridge magnets. Electromagnetism is at the heart of the everyday mechanics of the world.

Whenever two objects interact physically – when you push something or touch it or lift it or sit on it, for example – electromagnetism is the force that links the two objects. You might think when you push a button that your finger is touching the plastic. But in fact the electrons of the atoms in your fingertip repel the electrons in the atoms of the button. There is no contact. It is this electromagnetic repulsion that transfers your push to the button.

Similarly, on your theme park ride it is electromagnetism that is responsible for any contact between you and the carriage, or the carriage and the track. Of course there is also gravity at work. And thanks to the equivalence principle, we know why it felt as if you were much heavier as you cannoned round a bend, pushed hard against the side of the car. But electromagnetism is always present, acting between you and every object you are in contact with. Electromagnetism is everywhere.

Experiment – Gravity is a weakling

Take a fridge magnet and hold it at waist height away from any metal objects, then drop it. It's no surprise that it falls to the ground. Now hold it at the same

---

height, but very close to a fridge or other metal object. Drop it. It sticks to the fridge. Despite the whole vast Earth pulling down on it with gravity, the tiny object's magnetic attraction to the metal is enough to hold it up.

If you carried out that experiment, you might wonder what the point of the first part was – of course the magnet was going to fall to the ground. But this is where science differs from ordinary life. You can't assume what will happen. Common sense often lets us down when it comes to science. It's always best to test things out to make a meaningful comparison.

That example used magnetism, but you could also use electricity in a similar way, for example by picking up small pieces of paper with a comb you have rubbed on your hair to give it an electric charge. Electricity and magnetism are all part of the same force, a force that is vastly stronger than gravity. We're talking around 10^{40} times stronger – 1 with 40 zeroes after it. The only reason that gravity is so important is that atoms and molecules mostly don't have an overall charge (it's the charges on the subcomponents of atoms that come into play when objects touch), which leaves them neutral to electromagnetism, but still affected by gravity.

Going with the current

In your everyday life you can hardly avoid one aspect of electromagnetism – electricity. Electricity plays a fundamental role in keeping your body working. Your brain and nervous system, for example, use electrical impulses as part of the communication mechanism that controls your body's actions. Your heart's regular beat is activated by an electrical impulse.

Most of the lessons we get about electricity at school involve playing with batteries and lights and circuits, but you can do this to your heart's content and never really grasp what electricity *is*. In a sense this isn't too surprising; electricity, like pretty well all the 'workings' of physical science, operates at the counter-intuitive quantum level.

Electricity is often described using a model that pretends it's like a flow of water, but this is a really bad comparison. If electricity did run along wires like water down a pipe we would have to plug up electrical sockets to stop the electricity dripping out. Even so, thanks to the Victorian use of this model we have plenty of fluid-based terms associated with electricity, such as 'current' and the early electronic switching device, the valve (now replaced by solid state switches).

Electrical current works because conductors, such as metals, have loose electrons floating about, shared between the atoms in the substance. Let's say we put a positive charge on the right-hand end of a piece of metal – these negatively charged electrons would then be attracted towards it. But there's a problem. As all the electrons bunch up at the right-hand end, the left-hand end becomes short of electrons. Shortage of electrons means that the left-hand end of the metal now has a positive charge, pulling the electrons back again. But feed electrons into the left-hand end and the build up of positive charge is neutralised. So unlike water, electricity will only flow when there's a complete circuit, linking the ends.

It's rather unfortunate that the people who devised the model of electrical current didn't know about the existence of electrons. They made a totally arbitrary decision about which way current would flow, and it happens to be the opposite way to that of the true flow, the movement of the electrons.

The other problem with the water model is that it suggests that electrons pour down a 'tube' to provide the current. But if that were all that happened, we would have a long time to wait for electrical devices to kick into action. An electric light reacts pretty well instantaneously when the switch is thrown. Yet if you measure the speed of electrons down a wire, they saunter along at less than walking speed. (They actually shoot around at high speed, but all over the place – most of these movements cancel each other out, but add them all together and you get a gradual drift towards the positive pole.)

What is coming from the battery is not just a bunch of electrons, but an electromagnetic field – the field of influence of electromagnetic energy – and that travels at the speed of light. When you flick the switch it is this invisible wave (a stream of photons) that gets the electrons that are already at the lightbulb moving – they don't (thankfully) have to travel the entire length of the cable.

In fact, electromagnetism is involved in all interactions between light and matter. So it's not just the way that we touch things or run electrical devices. Without electromagnetism we wouldn't be able to see anything, nor would the energy from the Sun, crossing space as light, be able to heat up the Earth.

Into the nucleus

For completeness we ought to take a quick look at the other two forces that work alongside gravity and electromagnetism. They are important to your existence and the functioning of your body, but they aren't so immediately obvious. The more powerful of the two is rather unimaginatively called the 'strong nuclear force'. This one beats even electromagnetism. This is just as well, as without it, all the atoms in your body would ping apart into their individual components.

In an atom's nucleus it is the strong force that keeps the positively charge protons from flying apart. The electromagnetic force wants them to get as far away as possible, but the strong force overcomes this, holding the nucleus in a tight bundle. If it weren't for the strong force, every atom in your body would fly apart.

If the strong force only fell off with the inverse square of distance, like gravity and electromagnetism, we would be doomed. Every nucleus in the universe would be unstoppably attracted to every other. But the strong force drops in strength much, much more quickly. By the time something is around 10^{-15} metres away from a proton or neutron, the strong force is practically zero. This is why you don't get truly enormous atoms. Anything with a nucleus bigger than uranium has trouble staying together.

That's only half the story, though. The strong force that keeps the nucleus together is a kind of side effect; the result of the force leaking from its most dramatic role, which is keeping quarks where they belong. Every proton or neutron is made up of three separate quarks, and the strong force stops those from escaping. Unlike any other force, at the range quarks exist in, the strong force doesn't get weaker as they get further apart but stronger. Within a proton or neutron the quarks move freely, but if they try to separate the force gets intensely powerful very quickly. It's pretty well impossible to break a proton or neutron into its component parts.

The close-up force

By comparison, the fourth force is an oddity. This, the 'weak nuclear force', is around a million times weaker than the strong force (overcome by electromagnetism, though still beating gravity to a pulp). It isn't a simple attraction or repulsion between particles – even shorter range than the strong force, this weak interaction requires particles to be a tiny fraction of the diameter of a proton away from each other to exert itself.

The weak force acts as a switch for quarks, changing them from one 'flavour' to another – the result is that nuclear particles can change type, as when a proton switches into a neutron in the nuclear fusion reactions in a star, or during nuclear decay processes like beta decay which pumps high-energy electrons out of the nucleus. So even though the weak force isn't exactly essential to your rollercoaster ride on the face of things, without it the Sun would not be burning and there would be no life on Earth. In fact there would be no Earth – because the nuclear reactions in stars would never have made the heavier elements.

With all these forces at work on you during a roller-coaster ride, it's no wonder that when you get off you can feel a little battered and lightheaded. But do you also feel more youthful? The fact is, as a result of taking that ride you are now a fraction of a second younger than you would be otherwise.

Travelling through time

Let's take a more extreme example. Imagine you volunteered to take a ride in a spaceship, a new design that could fly at 99 per cent of the speed of light. That's not a trivial speed – 297,000 kilometres per second – but this is an imaginary flight, after all. You fly off into space for a round trip that takes around two years and nine months. When you get back, you get something of a shock, though. While you have been away, twenty years will have gone by on Earth. All your friends and all your family will be twenty years older. Think what has happened in world events in the last twenty years – imagine you missed all that. In effect by taking that trip you have time travelled over seventeen years into the future.

Both the time-travelling space trip and your tiny reduction in aging on the rollercoaster are down to one of the most revolutionary bits of science of the twentieth century: Einstein's special relativity. Einstein realised that there was something special about light. It can only go at a particular speed – around 300,000 kilometres per second – in a vacuum.

This is because light is a special interaction between electricity and magnetism. Move a source of electricity and it makes magnetism. Move a magnet and it makes electricity.

Get an electrical impulse moving at just the right speed – the speed of light – and electricity makes magnetism makes electricity and so on. A photon of light flies along constantly remaking itself. But this process can only happen at that exact speed. Slow it down at all and it stops working.

Anything else has a speed that varies depending on how you move with respect to it –from the queue, that theme park ride might shoot past at 60 miles per hour, but on board the ride, the carriage doesn't move with respect to your body (apart from jiggling about). Instead it's the scenery that flashes past – usually, all motion is relative. If two cars collide head-on, each travelling at 60 miles per hour, the resultant crash is at 120 miles per hour. But light is different. It doesn't matter how you move towards it or away from it, it always goes the same speed.

Light gets relative

When Einstein put the fixed speed of light into the simple rules of motion that had been around since Newton's day, something had to give. Things that had once been unvarying – the mass of an object, or the rate time flowed – had to shift. As you move faster and faster, time slows down, your mass increases and your length decreases in the direction you are moving. This is special relativity in action.

What special relativity also says is that normally it's not possible to move faster than light. Time gets slower and slower until it comes to a standstill at light speed. If it were possible to go faster, you would be able to travel backwards in time. Despite this apparent limit, though, there are ways around the light-speed barrier.

The simplest way to get something moving faster than light (though you can't use it as a time machine) is happening right now in every water-cooled nuclear reactor around the world. If you could see the water that surrounded the reactor core, it would be filled with an eerie blue light. This glow is produced by electrons travelling faster than light.

As we've already seen (page 85), light travels slower in water than it does in air (and slower in air than it does in a vacuum). The ultimate speed limit, the barrier beyond which time could be reversed, is the speed of light in a vacuum. But things can, and do, travel faster than the speed of light in water, which is around 225,000 kilometres per second. The electrons being pumped out in the nuclear reactor (produced by the weak nuclear force) travel faster than this.

As the electrons shoot past the molecules of water, they disrupt other electrons, blasting out light energy in what's known as Cherenkov radiation. It's sometimes likened to the sonic boom that a plane makes when it travels faster than the speed of sound – that blue glow is an optical boom produced by the faster-than-light electrons.

Tunnelling through time

Another way to move faster than light is to use the quantum tunnelling we met on page 73. When a quantum particle jumps through a barrier, as it does to fuel the Sun, it doesn't travel through the space in between. It takes no time to get from one side of the barrier to the other.

This means that if it's traveling from A to B including a section involving tunnelling, overall it will have moved faster than light.

Experiment – Faster than light Mozart
Go to **www.universeinsideyou.com**, click on *Experiments* and select the *Faster than light Mozart* experiment. Click on the sound player at the bottom of the page to play a signal that has been sent through a tunneling barrier averaging 4.7 times the speed of light over its journey. There is a lot of hiss, but the signal is clearly distinguishable.

In principle anything travelling faster than light, including a signal sent through such a barrier, is travelling backwards in time. But the further the particle has to tunnel, the less likely it is to get through. The phenomenon has only been observed over jumps so short that by the time the signal is read, any slip backwards in time is more than lost, so we can't send the lottery results back in time this way.

Build your own time machine

Your body is constantly travelling forward in time whenever it moves relative to anything else. Amazingly, though, in the future we could conceivably have access to time machines that could travel into the past. Unlikely though time travel seems, there is no physical law that prevents it. Travel into the past is more difficult than the future – certainly well beyond our current technology – but not physically impossible.

A theoretical physicist will tell you it's just a matter of engineering. All you need is to make a wormhole – a tear in reality that links two points in space-time – keep it open with antigravity and fly through it. Or take a string of neutron stars, form them into a cylinder and spin them at near the speed of light. Fly around the cylinder and you've a time tunnel into the past. These are feats that are millions of years beyond today's technology, but there is one possibility that could create the same effect as those spinning neutron stars.

The process relies on something called frame dragging. One of the minor details of general relativity is that there is a small component of gravity sideways to the normal direction of pull. When the body that is causing the gravitational pull is spinning around, that sideways pull drags space-time with it, pulling it like a spoon pulls treacle with it when you turn the spoon in a pot of it. Drag the space-time fast enough and it will produce a space-time vortex that makes it possible to travel backwards in time.

A US physicist has proposed constructing such a time tunnel out of lasers, where the spinning body is replaced by light itself. There are some technical issues with this device being constructed, but at the time of writing, funding is being looked for to turn it into reality. The first version would only allow small particles to drop back slightly in time, but unlike the quantum tunnelling, if this does work it could be blown up to a larger scale and achieve real trips backwards in time.

Before anyone plans a journey to meet a favourite character in history, this time machine has the same limit as any approach based on relativity. You can never travel further back in time than the point when the machine was first created. So no one could use it go dinosaur hunting. But it would still produce strange paradoxes.

The paradoxes of time

The most famous possible outcome of travelling backwards in time is that someone could go back in time to before their own birth and kill one of their parents or grandparents. (You can't do this, as you have been alive before the building of a time machine but someone born after the construction of one could.) I'm not sure why anyone would want to do this, but they would get into a paradoxical mess if they tried, because if they killed their parent, they couldn't be born, so they couldn't kill their parent.

Some people think such paradoxes prove that travelling backwards in time will never be possible. But it could be that the effect of generating such a paradox would either be to bounce the time traveller into an alternative universe, where their parent was still alive, or to flip them back to the point where they first travelled into the past, making the paradoxical action cancel itself out.

Here's another strange possibility: get hold of a copy of a book that has been written since the time machine was built. Take that book back in time and give it to the author before they wrote the book. They copy the text and submit it to their publisher. Now who wrote the book? It wasn't the author – he or she just copied it from the printed version. The book sprang into existence of its own accord. Mind-boggling, but possible if time travel ever becomes feasible.

Breathing easier at the theme park

Back in the theme park queue, if you suffer from asthma this could be a good place to be for your health. In an unusual piece of research in Holland, 25 young women with severe asthma (and fifteen control individuals who didn't suffer) were sent on repeated rollercoaster rides. It was discovered that the asthma sufferers found themselves less short of breath after the rides, even if they had a form of asthma where the motion of the rollercoaster reduced their lung function.

The conclusion drawn was that positive emotional stress (that 'Whoo-hoo!' lift you feel at the end of a rollercoaster ride) reduced the perception of being short of breath, while negative stress made the asthma symptoms worse. Apart from being a surprising outcome, it seems asthma sufferers would benefit from being more thrill-seeking than their typical image suggests, which leads rather nicely to the next chapter and the greatest everyday source of highs and lows.

7. Two by two

At any age, meeting an attractive person of the opposite sex can turn many of us into gibbering idiots. It doesn't seem a problem for other animals. Outside of the heat of the moment, they just get on with life, but we seem to lose our ability to think or control our body. What's going on here?

What do you mean, attractive?

It's worth getting a better feel for what makes someone else attractive before worrying too much about why it interferes with our brains. By 'attraction', I am primarily considering physical appearance. This may seem very shallow, but in a sense it's the reverse. Of course we find plenty of other things interesting in potential partners – conversation, wit, intelligence, personality – but these are about being a good companion. As far as your body is concerned, biologically speaking, attraction is all about the ability to reproduce well. This is fundamentally what *attraction* is – all those other good things are companionability.

So what makes someone attractive? There are a number of key factors, including:

* Youth – it doesn't matter how young or old you are, youth in another person (provided they have reached maturity) means they are more likely to be able to reproduce successfully.

- Healthiness – an essential attribute when thinking of the biological 'value' of a potential partner.
- Symmetry – we are attracted to other people who have symmetrical bodies, particularly symmetrical faces. Plenty of experiments have been done bring this out, where small changes have been made to photographs. It is probably because asymmetry is often linked to bad health.
- Approachability – for fairly obvious reasons, if the aim of attraction is reproduction, we value being appreciated back, as it implies that things should progress without aggression.

There is an interesting experiment that emphasises the contribution of mutual appreciation to attractiveness. If you show people photographs of faces, and doctor some of the photographs to make the pupils of the eyes bigger, these photographs will seem noticeably more attractive than the same image with smaller pupils. This is because your pupils dilate when you find someone else attractive – it's an involuntary response. So when you see the image of someone else with dilated pupils, your brain assumes that they are interested in *you*, and so you find them more attractive.

Birds do it, bees do it ...

Without doubt the most bizarre experiment ever to study human attractiveness (and there have been plenty of bizarre ones) involved chickens. Researchers at Stockholm University had trained chickens to select male or female human faces. It was discovered that these chickens then exhibited a preference for faces that would generally be regarded as more attractive by humans. While this was a limited experiment, and certainly not definitive, it seems to suggest that the same, basic qualities of attractiveness are recognisable even by non-human observers.

Attraction is, of course, rather different from falling in love – but this process too has been subjected to scientific testing. Many people go through a period of unusual behaviour when they first fall in love, and tests on the protein that carries the neurotransmitter serotonin to the brain showed that people who have recently fallen in love have a consistently unusual pattern in the sites that accept the neurotransmitter, implying that the chemistry of the brain is altered in a way that is similar to that of people suffering from obsessive-compulsive disorder (OCD).

Although attraction and the act of bonding as a pair have a whole host of implications, the biological imperative underlying them is to reproduce. We have a tendency in modern society to push this to one side, because it's something we have no choice about, and we don't like the idea that our bodies are overriding our brains. But there can be no doubt that your behaviour is strongly influenced by the parallel natural requirements of improving your survival chances and passing on your genetic material by reproduction.

Sometimes you will see this presented in a bizarre extreme that considers the genes to be in charge, with their goal being to ensure that they are passed on, hence the idea of 'the selfish gene'. But this is the biologists' equivalent of the way physicists tend to ignore friction when thinking about moving bodies, or simplify complex shapes as spheres. It doesn't provide a complete picture of human behaviour and fulfilment, though equally it would be blinkered to suggest that the sexual urge, driven by the need to reproduce, does not lie behind a lot of our behaviour. We have a strong natural instinct to create new life.

You can't make an omelette without breaking eggs

In your case, just like a chicken, life started off with an egg. Not a chunky thing in a shell, laid in a nest, but an egg nonetheless. But there is a significant difference between a human egg and a chicken egg that has a surprising effect on your age.

Human eggs are tiny. They are, after all, just a single cell, and are typically around 0.2 millimetres across. That's not too dissimilar from the size of a printed full stop. The egg that you came from was formed in your mother, but the surprising thing is that it was formed when she was an embryo. The formation of that egg, and so the half of your DNA that came from your mother, could perhaps be considered as the very first moment of your existence. And it happened not the length of your lifetime ago, but the sum of your age and your mother's age when you were born. Say your mother was 30 when she had you, then on your 18th birthday you could say that you were 48 years old.

Doing it the prehistoric way

This is quite an abstract idea of your beginning, though. We tend to think of our real beginning as our emergence as a separate, living entity at birth. If, like me, you are over 50, the chances are quite high that you were born around 2 a.m. (I was.) These days many births are sufficiently controlled that this is less likely to happen, but there seems to be a natural tendency for births to occur in the quietest part of the night. In a study of zoo chimpanzees, around 90 per cent were born in the middle of the night, not long after midnight.

It seems likely that we have inherited the tendency to give birth at these 'inconvenient' times because they are the safest for a potential prey animal to be born. Until we developed technology, human beings were more prey than predator. At the time of birth both baby and mother are particularly defenceless, and so benefit from having everyone else around, rather than out gathering food as they would be in a hunter-gatherer society during the day.

This is one of many examples of behaviours and responses we have that are better suited to when humans first came into existence than they are to the present. We have not significantly evolved in 100,000 years, and biologically – including in the way our brain acts, as we will see in the next chapter – we are evolved to deal with the kind of world that existed back then, not the world we now inhabit. We are still more scared of snakes than cars, even though 1.25 million people a year are killed on the road as opposed to only tens of thousands by snake bites (and a tiny percentage of those in Europe or the US).

Instead, the big changes that have been made to humans since *Homo sapiens* evolved have tended to come from our brains, and from the way we have

developed technology. Probably the biggest early transformation was our move from being prey animals to the
ultimate predators. Almost all our very earliest technology – using a lump of rock in the fist, for example – was
about driving this change of role.

The Stone Age technology in the park

Arguably, then, what makes you and your body most different from every other animal on the planet is your use
of technology in the broadest sense. We might make a fuss
about Stonehenge as the sophisticated peak of ancient
technology, but if you take a walk in any park you are
likely to encounter a much older piece of technology that
is still in use today and that made a big contribution to
the success of our move from prey to predator – the dog.

This may seem a little bizarre. How can a dog be a
piece of technology? It's a living creature. Yet dogs have
two distinct differences from wolves, the wild animals
they were bred from, that make them produced rather
than natural. First, dogs have functions – they don't just
exist alongside human beings, but carry out activity on
our behalf. And secondly, dogs were the first example of
animals with deliberate genetic modification, bred with
a particular intent in mind.

A dog can run faster than a human being. It has a much
more effective sense of smell. Its jaws are more powerful,
and its fangs larger and more dangerous than a human's
comparatively weak teeth. If you consider a hunting
and protection dog – the two initial roles of 'man's best
friend' that helped us become effective predators – it
makes a formidable weapon that can work when we are

out of sight, and presents a confusing second source of danger for any attacker.

Because of their pack loyalty, dogs rapidly became more than tools, developing a close and complex relationship with their owners. That the relationship is complex can be seen in the way attitudes to the dog have changed with time, and in different cultures. Though practically every civilisation has made use of dogs, there have been widely differing views of their nature. In Middle Eastern cultures, dogs are often viewed as dirty scavengers, and a lot of our invective involving dogs, inspired by Biblical language, still labels them dirty, lazy, greedy and shameless.

This didn't stop dogs being used in profusion, and by the late middle ages, a strong distinction was growing between the 'noble' hounds kept by the aristocracy and allowed freedom of the home, and working dogs, treated with as little care as any other animals in the period. The distinction between pets and working dogs is maintained to some extent to this day, though it is no longer reflected in a separation of breeds, as practically every type of dog is now kept as a pet.

Historically the breeds were selected on the basis of traits that made them suited for a particular role. Heavy-set mastiffs as guard dogs and hunting dogs; intelligent, gentle retrievers to search out and bring back fallen prey; wiry terriers to go down fox holes or to take on rats; sensitive hounds to follow scent – like any flexible piece of technology, the dog was developed into many different models to suit varying needs

Some of those uses are still with us today. Although the majority of dogs are now pets, working dogs still extend human capability, some in ways that couldn't have been dreamed of when dogs were first bred. After hunting and protection, dogs came to be used to pull small carts and sleds, to turn spits in the fireplace and to track down criminals. On the farm, the dog became indispensable as a patient assistant in rounding up sheep. The hunting dog breeds diversified – no longer were they just assistant killers, but split into hounds, pointers and retrievers.

Dog as prosthetic

Most remarkable of all is the role that dogs have fulfilled as extensions to the human body by being helpers to the blind, the deaf and the disabled. There is some evidence of dogs being employed to help blind people early in history. When the Italian town of Herculaneum – buried beneath the ashes of the volcano Mount Vesuvius when it erupted in 79AD – was excavated one of the murals found featured a blind person being led by a dog, while a medieval wooden plaque also shows a blind man being assisted by a canine helper.

The concept was mentioned in passing in a couple of nineteenth-century books, but no one seems to have taken it seriously until the First World War. The earliest organised attempt to train guide dogs was in Germany in 1916, when they were intended to guide soldiers who had been blinded in battle. This idea spread to America in 1927, when an American woman working as a dog trainer in Switzerland, Mrs Dorothy Eustis, found out about the German work and wrote an article that was picked up by the first American owner of a 'seeing-eye dog', Morris Frank, and his dog Buddy.

Since then, thousands of people have been able to recover an active life thanks to guide dogs. I recently watched a guide dog lead its owner from a train to the exit of Paddington Station in London. Despite the milling crowds, ticket barriers, a 'wet floor' warning sign and a whole host of hazards that seemed to have been put in the way deliberately to make the task of crossing the station difficult, despite the noise, the smells from Burger King and Krispy Kreme outlets, and the nearby presence of the huge, noisy trains, the dog was able to lead its blind owner at normal speed across the station and on his journey.

More recently, guide dogs have been joined by other types of assistance dog. Hearing dogs alert their deaf owners to audible signals that a hearing person would pick up and respond to – it might be a ringing doorbell, or the sound of a reversing vehicle nearby. Although a hearing dog doesn't need the same precision as a guide dog, it has to make sophisticated distinctions in the melee of sounds that makes up modern life.

The third class of assistance dog is a service dog, trained to help those with physical disabilities that make it difficult to be mobile or to manipulate objects. It is quite remarkable to see one of these dogs operating an ATM on behalf of its owner.

Genetic engineering the natural way

Of course the production of this remarkable piece of technology to assist the capabilities of the human body didn't begin with the intention of creating such a flexible helper. The chances are it all started by accident.

Although wolves don't deserve a lot of the bad press they get – they rarely attack human beings, for instance – they would have been irritating scavengers that early man had to make an effort to see off, to stop them stealing the remains of hunted animals.

It's easy to imagine those first, tentative steps away from the wolf's role as enemy. Perhaps it was a cold winter, and a wolf crept close to a fire to keep warm. Maybe while it was there some other predator attacked the camp – the wolf, ever the pack animal, jumped to the defence of the humans, fighting alongside them. It was rewarded with a gift of meat. Natural selection takes things forward from here. Over the years, wolf cubs that are more docile, that fit in more easily with a human pack, are the ones likely to stay around and to be fed and encouraged. Over tens or hundreds of years, the dog emerges.

Remember the experiment by Dmitri Belyaev mentioned in Chapter 2? In just 40 years he turned wild silver foxes into domesticated creatures something like dogs – the process really doesn't have to take long. Perhaps 100 years after that first tentative contact, the early hunters were no longer dealing with wild wolves. The animals that lay around their camp had changed in manner and appearance: their once upright ears had drooped, their coats were more varied in colour and they accepted humans as part of their pack. The dog had been created.

This was genetic engineering, just as much as any GM crop. By selecting for certain traits, humans have modified the nature of many animals and plants to better suit their requirements. This is particularly obvious in two plants, cauliflower and sweetcorn (maize).

The cauliflower is a mutant cabbage – its flower has been transformed into a hard, bumpy white structure, the part we now eat. With no functional flower, it can't breed without help. Similarly, sweetcorn has been selected over the years for bigger and bigger seed husks. It is now incapable of self-seeding and won't grow without human assistance.

Just as these plants are no longer viable in the wild, the dog is not a natural animal. It is as a much a human-made piece of technology as a table that started off as a natural piece of wood. Without doubt, the dog is one of the most impressive early technologies we used to enhance our lives. Forget Stonehenge, it's a toy by comparison. Okay, it gives a handful people some astronomical information, and it's pretty, but it hasn't been used for thousands of years. The dog is a piece of Stone Age technology, developed 35,000 years *before* Stonehenge to enable our ancestors to go beyond the limits of the human body, and it's still going strong.

The mighty 23

As we have seen, every living thing, from those dogs, to you, to what you ate for breakfast, is constructed according to the 'control program' in its DNA. It's time to take a closer look at this remarkable set of chemicals and their role in your body. We've already heard that each human being (with the exception of a minority of people suffering from genetic disorders) has 23 pairs of chromosomes, each containing a single molecule of DNA. These come in matching pairs in every one of your cells (with the exception of number 23, where things get more complex).

This pairing of chromosomes reflects your origin from two parents, one chromosome in each pair coming from the mother and one from the father. This might seem unnecessary overkill, but the new version of each chromosome that comes from you in the embryo will be made up of a mix of bits from your own chromosome pairs, ensuring genetic diversity in the way the human race continues.

Although each pair of chromosomes contains the same genes (except for number 23, on which more in a moment), you do need a set from each of your parents. If eggs are made with two sets of chromosomes from a male or two from a female, the cells don't develop properly. This demonstrates the importance of epigenetics (see page 218), the science that goes beyond the information in the genes. The external factors that influence how the genes operate differ in the versions from mum and dad, and those differences are essential for healthy development.

Chromosome number 23 is the odd one out because this is where the big variation between men and women occurs. If you are a woman your chromosome 23 pair are of the same design (each a so-called X chromosome), but male readers will have one X chromosome from their mother while the other half of the pair will be a much smaller Y chromosome from their father.

Each chromosome contains a really long DNA molecule, and this is where that structure we discovered in Chapter 3 is so important. Stretched out, DNA is a bit like a spiral staircase, with each tread of the staircase having one of four possible bases – cytosine, guanine, adenine and thymine – on one side, and a matching base on the other. Your genes, those much talked about components of life, aren't separate entities; they are just segments of the DNA molecule.

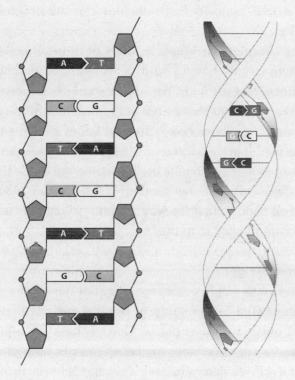

A segment of DNA spiral with a pop-out showing
the CGG coding for arginine

A gene is a collection of triple 'treads' of the DNA staircase, so it will consist of various three letter 'words' in the DNA code (we use the initial letters of the four bases to identify them). One word, for example, might be CGG (cytosine, guanine, guanine). These letter combinations

are at the heart of the way genes work. Each triple combination of the four possible letters identifies a specific
type of chemical called an amino acid, or else is a control
code instructing the mechanism for reading DNA to stop.
So, that CGG code, for instance, indicates the amino acid
arginine.

The complete gene uses a series of these three-letter
words to specify how to build a protein, the workhorse
chemicals of your body. You have somewhere between
20,000 and 25,000 genes – not very many, really, to specify everything about how a human being works, so it's
just as well that this isn't a role that is left to genes alone.
At one point, if you read a biology book you'd think that
genes were all that was needed, but since the 1980s it
has been realised that the way your body is built is much
more complicated than this.

Beyond the gene

The secret lies in two concepts that fall within the field
of 'epigenetics' – the study of the instructions that are
coded outside of your genes. One of these concepts is
that genes are not always operating but can be switched
on and off. A common way this happens is through
methylation. This involves sticking an extra collection of
molecules known as a methyl group (just a carbon atom
with three attached hydrogens) onto one of the bases that
form the treads of the DNA spiral staircase. These little
molecular blobs act as markers to control the way a gene
is used – or ignored.

The other thing you need to know to gain a better
understanding of how your body built itself lies in those

huge molecules of DNA. When you hear that you have many fewer genes than, say, a rice plant, it sounds a bit humbling, and genes certainly have an important job in specifying the proteins that your body makes. But your genes are only a tiny part of your DNA; around three per cent, to be precise. The other 97 per cent was originally thought to be rubbish – 'junk DNA', left over from past evolutionary stages. But this couldn't be more wrong.

Much of that 'extra' DNA has very important functions. A lot of it, instead of specifying how to build proteins, specifies how to make RNA. This is a compound that is related to DNA but only has a single backbone strand. RNA is used in the process of building proteins from genes. In effect, the control program of the gene produces a mould of RNA in which the protein is built – the RNA acts as a kind of messenger.

It used to be thought that the RNA produced by the 'junk' DNA was just useless historical baggage, but it turns out that this RNA is valuable in its own right. It provides many of the control mechanisms for switching genes on and off, as well as playing other roles that can be just as important as the way proteins are used. Suddenly, what was a relatively small program in just those 20,000 plus genes has become an enormous one, where all of your DNA has to be taken into account.

The message here is that it's easy to read too much into genes. Epigenetics demonstrates how it isn't simply a case of genes providing a blueprint for human beings. Yet spurred on by the image of 'the selfish gene' ruling living things, a concept encouraged by Richard Dawkins' famous book of that name, it's all too easy to give genes too much emphasis. Dawkins' book was written before the true significance of epigenetics was realised (he has since added a chapter on it). It's not that genes have lost their importance, but we now realise that they are a relatively small part of the whole biological control program.

Similarities and differences

You will often see it said that we are, genetically, very close to chimpanzees. And it's true that our genes are surprisingly similar. Around a third of the proteins produced from them are identical, and most of the rest only differ by one or two of the base pair codes – they have a couple of differing amino acids, but are basically the same. However, there is much bigger variation in the rest of the DNA that doesn't code for proteins.

One big difference is in the way we modify the RNA molecules produced by those sections of DNA that don't code for proteins. There are various ways these molecules can be changed after they are produced, a process known as editing. Humans edit this non-coding RNA more than any other species, even our ape cousins. And this process happens in the brain more than anywhere else. This could be an explanation as to why our brains are functionally so different from animals that we are very closely related to genetically.

There's another oddity about those genes. Scientists at the University of Michigan in Ann Arbor compared 14,000 matching genes in the human and chimpanzee genomes. Of these, 233 of the chimp genes have changed as a result of positive selection – where natural selection appears to have kept a change that gives benefit to the species – as opposed to just 154 of the human genes. The lead researcher commented: 'The result overturns the view that, to promote humans to our current position as the dominant animal on the planet, we must have encountered considerable positive selection,' while a primatologist, Victoria Horner, said 'We assume chimpanzees have changed less than us when that's actually not the case.'

From the outside it's hard to see how biologists could have such a blinkered view. Clearly we have changed far more from the first *Homo sapiens* than chimps have from early chimps. The distorted view of the scientists is arguably all the fault of physicist Ernest Rutherford, the man who discovered the structure of the atom. Rutherford once said 'All science is either physics or stamp collecting.' What he meant was that physics had the explanatory insights while other areas of science, particularly biology, are almost all about cataloguing what is out there.

There was an element of truth in this until biology came up with evolution and genetics, which transformed the science. This is probably why biologists, irritated by comments like Rutherford's, sometimes give too much weight to the genes. Sheer numbers of genes don't give a useful picture of the complexity of an animal or plant. Even the brightest rice plant, with all those extra genes compared to a human, is not likely to write any great literature, make a scientific discovery, or have any exciting plans for the future. Epigenetics ensure that a small number of genes can be responsible for a phenomenally important difference in a creature, with our large brains as a prime example.

To use genetic change alone to suggest that chimpanzees have changed more than human beings is a perverse focus on one small aspect of a totality. We aren't just our genes – thanks to our remarkable brains many of the changes that we have produced are down to technology and the way we interact with the world around us. To say that in the last six million years, chimpanzees have changed more than human beings is ludicrous.

In that time chimps have, well, carried on as usual – they have kept on doing what chimps do with very minor changes. They haven't developed the ability to fly. They can't cross deserts with no water holes for days and live. They can't exist in space (unless we make it possible for them). They can't survive illnesses that should kill them, or see what is happening on the other side of the world. Our quasi-evolution through the capabilities of our brains leaves the chimpanzee on the evolutionary starting block.

Attack of the clones

One of the most misused ideas to come out of genetics is that of cloning. According to Hollywood, if you want a more substantial copy of your body than the one you see in the mirror, you need a clone. This is the process of producing another creature with DNA identical to a single individual of the same species. When we think of clones it's easy to imagine them as identical copies, but this is far from correct. Despite the fact human cloning is not currently possible, we have plenty of examples of human clones brought up in the same environment, yet differing considerably.

This apparent contradiction is possible because the human clones that exist – you have almost certainly met some – are natural. They are identical twins. Even though they start with exactly the same DNA, because they are created by a single egg splitting into two, identical twins are clearly unique individuals by the time they are adults – they often don't even look truly identical any more.

Not only do identical twins have subtle differences in their environment as they grow up – they can't experience *exactly* the same life – but they will be truly biologically different. Firstly, our genetic code isn't one hundred per cent fixed at birth. Each of us will gradually accumulate changes. For example, when a cell splits, something that is happening all the time in your body, DNA is duplicated. Errors can occur in this process, resulting in very small changes to the genetic code. In this sense we are all mutants.

More significantly, genes are not operative all the time. As we have seen, they are switched on and off at various points in your life, controlled by external chemicals. This epigenetic side to your development can make huge differences, and the switching on and off of genes is, without doubt, influenced by the environment. The result of these influencing factors upon twins is two unique individuals, not clones that are identical copies.

The difference between clones and copies was proved with some irony (as far as names go, at least) when the first cloned cat was produced at Texas A & M University. It was called Copycat (Cc for short), but it proved anything but a carbon copy of its parent. Its parent was calico, while Cc was tabby and white. This seems to have been an epigenetic effect, as the host Cc was grown in was also a tabby. So there really is no point having your favourite pet cloned to keep it around. The clone is likely to be a very different animal.

Hello Dolly

When Dolly the sheep was cloned back in 1996 it seemed that it would only be a matter of time before someone would clone a human being. The ethics of doing this are debatable, but it's difficult to put the genie back in the bottle. There were even a couple of organisations that claimed they had already cloned a person, though they never produced any evidence of this. The chances are it never happened, because one of the lessons learned from Dolly is that cloning is a tough business.

What usually happens in human (and most other animal) reproduction is that half the genetic content of the new person comes from one parent and half from the other. To make a clone it's necessary to get all of the DNA from a single individual into an egg. In the case of Dolly, this DNA came from the mammary of a long-dead sheep (the cell was from a culture kept alive in the laboratory), hence this famed animal being named after singer Dolly Parton, whose own mammaries are rather noticeable. The DNA from that cell was injected into another sheep's unfertilised egg, which had first had its normal contents sucked out.

The egg was then given a tiny burst of electricity, Frankenstein fashion, to kick-start the process. It was finally implanted in a host mother, where it began to grow in the usual fashion, resulting in Dolly, a straightforward healthy-seeming lamb, being born. (Note, by the way, that the clone has to grow to maturity just like any other baby – you can't clone a fully formed animal, or human, overnight as some movies portray.)

That sounds a simple process, and that therefore human cloning should be just around the corner – but actually it isn't simple at all. First the researchers who produced Dolly had to get cells into just the right state, as they don't automatically begin to split and grow. They found that the best kind to use were cells that had already started splitting, but that were then 'paused' by removing nutrients – these were the easiest to get started splitting again. Even so, most attempts were failures.

Out of 276 cells that were started, only 29 activated, and of these only one – Dolly – survived. And even then things weren't necessarily as positive as they seemed. Dolly died young, around half the age of a typical sheep. Ian Wilmut, the scientist behind Dolly has suggested that this was because of a relatively common infection. But equally it's possible that Dolly died of old age in her youth.

This can happen when there's a problem with telomeres. These are little 'tags' on the ends of the chromosomes, the DNA molecules that contain our genes. Each time a DNA molecule splits because a cell is splitting, it loses one of these tags, a mechanism to prevent runaway cell growth. (Cancer cells have the telomeres switched off, losing this control.) Dolly's telomeres started off identical with those of her six-year-old parent, so it is possible that this could limit the lifespan of clones from older genetic sources, where some of the tags will already have been lost as the original creature grows and repairs itself.

Growing old gracefully

Ageing remains something of a mystery. We can identify some of the mechanisms that cause us to age, many of them tied into our biological past, when humans ceased to serve any useful purpose once they had reared their children. Yet anyone who doubts the benefits of modern science and technology can reflect on the way life expectancy is on the increase. In medieval Britain, life expectancy was around 30. By early modern times in the UK and US it was more like 50. Through the twentieth century it has grown to the extent that it is now around 80.

These figures taken in isolation can be misleading, though. There's the well-known split between men and women, so that at the time of writing there is about a five-year greater life expectancy for women. But also we shouldn't assume that these statistics mean that most medieval people lived for around 30 years and then died. A lot of the increase in life expectancy over the centuries was produced by reductions in infant mortality – the deaths of the very young lowered the figure for the average life expectancy considerably. If you attained adulthood you would likely make it well past that average – if you survived to 21 in 1500, for example, you could expect to live to around 70. Before modern medicine, two thirds of children died before they were four. It's a sobering thought that until the twentieth century, the majority of funerals were for children.

Clones are particularly prone to infant mortality.

The genes of a clone can be damaged easily in the process of manufacturing them. At the moment, producing a clone is a bit like a craftsman trying to repair a watch with a hammer and chisel. Occasionally he will get lucky, but more often the process will damage or destroy the original. Artificial clones frequently have genetic problems, with many embryos not surviving and those that do often suffering from serious defects. The risk is worse with monkeys than other mammals, and worse with apes than monkeys – it is quite possible it would never be possible to produce a human clone without making many damaged children as a by-product. The risks are simply too high for any respectable scientist to attempt human cloning.

This doesn't mean, though, that it isn't possible to safely clone individual human cells, a process that could produce major health benefits. One of the biggest problems with transplants, for example, is that the human immune system, designed to protect the body against invaders, attempts to destroy foreign cells, even if they are in a life-giving implant. If it becomes possible to construct organs by cloning a patient's own cells there will not be the same risk of rejection.

We have only been able to skim the surface of the attraction between human beings, its causes and the underlying genetics that provides the original driver for that attraction. It's easy to think of physical attraction as being something that belongs to your body alone, a purely visceral response. But that's a mistake. Like so much of the rest of your life, the impulse comes from the most distinctive and certainly the most complex part of your body: your brain.

8. Crowning glory

As we tour around your body, experiencing the associated wonders of science, we don't find a lot that is unique to human beings. There is nothing we have experienced in the body itself that couldn't be found working similarly in other animals. Your eyes, for example, are fine, but nothing special. Every capability possessed by the parts of your body we have met so far can be bettered by a different creature. But there is one bit of you that is special. Your brain.

What goes on inside your head

That unappetising looking lump of flesh in your skull, weighing in at around 1.5 kilograms (three pounds) is fiendishly complex. Inside it, there are around 85 billion of the key functional cells, neurons, some with many connections to others, making the number of connections at any one time around 1,000 trillion. And considering that it only amounts to one or two per cent of your bodyweight, your brain is a real drain on resources – of the 100 watts or so of energy your body generates (equivalent to a traditional light bulb), the brain hogs around twenty per cent.

Look at a picture of the brain from above and it appears to be a single lump of matter, not unlike an enormous pink walnut, but in fact it is almost entirely divided into two, with the halves of the brain joined at the back by a bundle of nerves called the *corpus callosum*. Some responsibilities are split between the two halves. The left side is largely responsible for the right side of your body, including the vision from your right eye, and vice versa.

There is a traditional view that the left side is the one that kicks in when you are being organised and structured. It is largely responsible for numbers, words and rationality. It prefers things sequenced and ordered. There's nothing it likes better than taking an analytical approach, working through something step by step in a linear fashion. The right side in this conception of the brain is much more touchy-feely. It takes the overview, a holistic approach to the world. It deals with imagery and art, colour and music. If you need to think spatially or deal with aesthetics, it's time to call on the right side.

At least, that's the simplistic view. However, when we're dealing with the brain, things are very rarely simple. In practice, though one side may dominate, both sides are involved to some degree in all these types of thinking. What is certainly true, though, is that the brain has two clear modes of operation that correspond to the attributes traditionally allocated to its two sides (and so labelled left- and right-brain thinking). This is why it can often be a real problem to come up with fresh ideas in a traditional business environment.

People will sit down to have a nice, structured, orderly meeting. Very logical and analytical. Before long, the right sides of their brains have shut down, leaving the participants with limited resources for creativity, as new ideas depend on making fresh linkages, and the ideal is to have both sides of the brain in action. This is why new ideas can often be inspired by music, taking a walk, looking at images, thinking spatially. It's a way of bringing the right side of the brain in to play.

Experiment – Feeling your brain

There is a simple way to experience the two halves of your brain in action. A technique called the Stroop effect allows you to experiment on your own brain (no surgery required) and feel the switch between the sides. Go to **www.universeinsideyou.com**, click on *Experiments* and select the experiment *Feeling your brain*, then follow the instructions.

The Stroop effect uses words and colours, each a responsibility of a different side of the brain. It doesn't matter how much you are instructed to concentrate on colours, in this experiment your brain sees words – handled primarily by the left side of the brain – and lets the right side, taking care of colours, pretty well shut down. When you suddenly have to make use of the right-hand side again, you can practically feel the gears grinding in your brain as it tries to catch up.

Brains weren't made for maths

We've already seen when looking at sight and hearing that it is easy to fool your brain. The human brain is absolutely great at many things. But it often struggles with tasks that have been added to our repertoire since brains evolved.

A good example of a role your brain just wasn't evolved to work with is arithmetic. Your computer at home would be hopeless at many things you do easily, but give it a task like finding the square root of 5,181,408,324 and it will have the answer before you've even scratched your head. (It's 71,982, of course.) This just isn't the kind of thing humans were evolved to do – maths doesn't come naturally.

Nowhere is this more obvious than when dealing with probability and statistics. Probability is involved in many of our everyday activities, and statistics are thrown around in the news and politics all the time, yet our brains, developed to deal with images and patterns, have a huge problem dealing with those manipulations of numbers and the impact of chance.

Let's take three examples where the nature of your brain's wiring is such that it gets confused by these incredibly useful numbers.

Open the door

In the 1960s, Canadian-born presenter Monty Hall was in charge of a US TV game show called *Let's Make a Deal*. The format of the show resulted in the kind of problem that is excellent at exposing our difficulties with probability.

Let's imagine you're through to the final stage of a TV game show like *Let's Make a Deal*. The host brings you to part of the set where there are three doors. Behind two of these doors is a goat (don't ask me why), while behind the third door is a car. You want to win the car but don't know which door it is behind. Still, you are asked to pick a door, so you do. There's a one in three chance you have picked the car, and a two in three chance you have picked a goat.

Now the host opens one of the doors you didn't pick and shows you a goat. He gives you a choice. Would you like to stick with the door you first chose, or switch to the other remaining door? What would you do? Does it matter, in terms of your chance of winning the car? Is it better to stick with the door you first chose, better to switch to the other unopened door, or does it not matter which of the two you choose?

We know that after one door is opened to show a goat there are two doors left, one with a car behind it, one with a goat behind it. So it seems obvious that there's a 50:50 chance of winning the car whichever door you choose. And yet this is wrong. In fact you would be twice as likely to win the car if you were to switch to the other door as you would if you were to stay with the one you first chose.

If you find that statement ridiculous, you are in good company. Writer Marilyn vos Savant had a column in *Parade* magazine in which she answered readers' questions. In 1990, she was presented with this problem and came up with the answer I gave above: you are better off switching, it's twice as good as sticking. She was deluged with thousands of complaints telling her that she was wrong and that there was an even chance of winning with either of the remaining doors. Some of the letters were from mathematicians and other academics.

You can easily demonstrate that it is better to switch using a computer simulation – it really does work. But that doesn't get around the frustration of it not seeming logical. The important factor is that the game show host didn't open a door at random. He knew that there was a goat behind the door he opened. Think back to when you first chose a door. There was a 2/3 chance you had picked a goat – a 2/3 chance that the car was behind one of the other doors. All the host did was show you which of those two doors to pick – there was still a 2/3 chance that the car was there. So with only one alternative, you were better switching to the third door.

The two-boy problem

Oddly enough, another of vos Savant's columns also created a surge of complaints, and this too was as a result of a problem with probability that strains the brain. The problem is simple enough: *I have two children. One is a boy born on a Tuesday. What is the probability I have two boys?* But to get a grip on this problem we need first to take a step back and look at a more basic problem. *I have two children. One is a boy. What is the probability I have two boys?*

A knee-jerk reaction to this is to think 'One's a boy – the other can either be a boy or a girl, so there's a 50:50 chance that the other is a boy. The probability that there are two boys is 50 per cent.'

Unfortunately that's wrong.

You can see why with this handy diagram. The first column is the older child. It might be a boy or a girl, the chance is 50:50. Then in each case we've a 50:50 chance of a boy or girl for the second child. So each of the combinations has a one in four (or 25 per cent) chance of occurring.

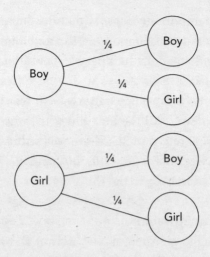

Potential combinations of children

All the combinations except girl-girl fit the statement 'I have two children. One is a boy.' So we've got three equally likely possibilities where one child is a boy, of which only one is two boys. So there's a one in three chance that there are two boys.

If this sounds surprising, it's because the statement 'one is a boy' doesn't tell us which of the two children it's referring to. If we say 'The eldest one is a boy', then our 'common sense' assessment of probability applies. If the eldest is a boy, there are only two options with equal probability – second child is a boy or second child is a girl. So it's 50:50.

Now we're equipped to move on to the full version of the problem. I have two children. One is a boy born on a Tuesday. What is the probability I have two boys? Your gut feeling probably says 'The extra information

provided about the day he's born on can't make any difference. It must still be a one in three chance that there
are two boys.' But startlingly, the probability is now 13 in
27 – pretty close to 50:50.

To explain this perhaps I should draw another diagram, but I can't be bothered – you'll have to imagine
it. In this diagram there are fourteen children in the
first column. 'First child a boy born on a Sunday, first
child a boy born on a Monday, first child a boy born on
a Tuesday, first child a girl born on a Sunday etc. all the
way through to first child a girl born on a Saturday.

Each of these fourteen first children has fourteen
second children options. Second boy born on a Sunday
... and so on.

That's 196 combinations in all, but luckily we can
eliminate most of them. We are only interested in combinations where one of the children is a boy born on a
Tuesday. So the combinations we are interested in are
the fourteen that spread out from 'first child a boy born
on a Tuesday' plus the thirteen that start from one of the
other first children and are linked to 'second child a boy
born on a Tuesday.' This makes 27 combinations in all.
How many of these involve two boys? Half of the first
fourteen do – one for a second boy born on each day of
the week. And for the remaining thirteen, six will have a
boy as the first child (because we don't include 'first boy
born on a Tuesday.') So that's 7+6 = 13, 13 combinations
that provide us with two boys. So the chance of there
being two boys is 13 in 27.

Common sense really revolts at this. By simply saying what day of the week a boy was born on, we increase

the probability of the other child being a boy. But we could have said any day of the week, so how can this possibly work? The only way I can think of to describe what's happening is to say that by limiting the boy we know about to being born on a certain day of the week, we cut out a lot of the options. We are, in effect, bringing the situation closer to being that 'the oldest child is a boy' – we are adding information to the picture.

The probabilities work – you can model this in a computer if you like – and the numbers are correct. But what is going on here mangles the mind. Don't you just love probability? (I ought to say, by the way, that this isn't quite realistic. It assumes there is an equal chance that either child is a boy or a girl, and that there are equal chances of children being born on each day of the week. In reality neither of these is quite true, but that doesn't matter for the purposes of the exercise.)

A test of your understanding

Those last two examples do come up in real life. As well as on Monty Hall's show, for example, a version of the goats and car problem was used on Mississippi river boats by gambling hustlers who got punters to bet based on the 50:50 assumption and made a killing. But the third example of how bad the brain is at dealing with probability and statistics is one that is much more important for real life, because it's one that rears its head in the way we understand the results of medical tests – and it's a difficulty that doctors have just as much of a problem with as the rest of us.

Let's imagine there's a test for a particular disease that gets the answer correct 95 per cent of the time, so it's quite a good test. Let's say that around one in 1,000 people – which would be around 61,000 people in the UK – have this disease at any one time. And finally a million randomly selected people take the test, including you. If you are told your test came out positive, how likely are you to have the disease?

Bearing in mind that the test is 95 per cent accurate, you may well think that you have a 95 per cent chance of having the disease, but actually the result is much more encouraging. Of those million people tested, around 1,000 will have the disease. Of these, 950 will be told correctly that they have it and 50 won't, as the test is 95 per cent accurate. 999,000 won't have the disease. Of these, 949,050 will get a (correct) negative result from the test and 49,500 will get a false positive result.

This means that of the 50,450 positive results, 98 per cent will be false. If you get a positive result, there is only a two per cent chance you have the disease. This example might use extreme numbers, but whenever you have a widely used test for a relatively rare condition, the chances are that the majority of the positive results will be false. This can be both distressing and result in potentially dangerous further testing, so it isn't a trivial outcome. Once more, the way our brains are made simply doesn't fit well with understanding probability.

But what does it mean?

Whenever your brain encounters probability and statistics, it's worth just taking a step back and making sure you understand what's going on. And make sure also that other people using statistics have got it right. It's all too common for government departments, newspapers and TV news desks to make just the same mistakes with probability and statistics as the rest of us.

A good way of testing statistics is to explore a little more widely – get some more information before you believe scary-sounding numbers. You might hear, for example, that violent crimes in your neighbourhood have increased by 100 per cent since last year. It sounds like it's time to move out. But make sure you ask for numbers to put this into context – if the rise is from one crime to two, it *is* a 100 per cent increase, but the reality isn't as worrying as the statistics sound.

You also need to be particularly careful to keep your brain on track when you have to deal with multiple sensory inputs. A great example of this was research conducted in the late 1990s where people were stopped in the street and asked to give directions. While they were helping someone with a map, some workmen came along the street, carrying a door. The workmen passed between the test subject and the person asking for help, who was one of the researchers.

While the door blocked the subject's view, the person asking for help swapped places with one of the door carriers. Around 50 per cent of subjects never noticed that they finished off giving directions to a totally different person. They were too focused on the task. We are much less conscious of what's going on around us than is often assumed in a court of law.

You must remember this

Memory is equally worryingly faulty. You are, in many ways, your memories; without them you would not be the person you are. Yet a fair number of those memories you cherish are false. Some are constructed a long time after the event to which they refer. It's not uncommon for what seems to be a memory to be derived from a photograph or video of an event. Others are slanted by our opinion – for example, we tend to remember extremes, so we think a summer was much hotter than it really was because of one hot day. We are also more inclined to give weight to recent experiences, so a wet week at the end of a month of excellent weather will have us moaning about not getting a summer at all.

Another problem with memories is that they are based on your ability to observe and capture information, but as we've seen, the image your brain shows you is a very subjective construct. This can easily lead to your seeing things that aren't there, or not seeing things that are, and these mistakes are subsequently remembered as fact.

A while ago, someone mentioned they had seen me walking the dog while I chatted on my mobile phone – quite a detailed observation. The only problem was, I wasn't at home that day, and hadn't taken my dog for a walk. This is where the whole business of observation, perception and memory becomes potentially dangerous. Imagine that the person who thought he saw me then witnessed a murder, committed by the person he saw. He would have been happy to stand up in court and swear that he saw me commit that crime, yet I wasn't there. Whenever a court case depends solely on witness evidence, particularly evidence depending on memory after a significant period of time has elapsed, it's quite worrying.

Experiment – Counting the passes

This is a very well-known experiment, but please have a go at it even if you have seen the original version – this is a new version that will still be of interest if you carry it through to the end. Go to **www.universe insideyou.com**, click on *Experiments* and select the *Counting the passes* experiment. You will be asked to count the number of times someone in white passes the ball. In the fast-moving game it is difficult to keep track (numbers and memory involved here), so you really need to concentrate hard on who is passing the ball.

Although it doesn't work for everyone, more than 50 per cent of people fail to accurately observe what is going on in this simple video. It's hardly surprising how often your brain will get things wrong. Often these failings are more entertaining than worrying – optical illusions can be great fun, for example. However, whenever we rely on our ability to recall exactly what happened in confusing circumstances, we need to be aware of the brain's limitations.

Memory lets us down in surprising ways. We might recognise a face – so clearly it is stored away in our memory – but be unable to put a name to it. It is entirely possible to forget your own phone number, even though it is a sequence of digits that you make use of time and again. Perhaps most frustrating of all is the way that memory can give you half the story – there are times when you know there was something you had to remember, but you can't remember what it is!

Solid state versus squishy state

One of the reasons it's easy to misunderstand memory is that we are so familiar with computers, and we assume that there is some similarity between the way computer memory works and the way human memory works. But that's not the case.

Computer memory consists of a specific value – zero or one – stored in a specific location. Each location has an address. You can go straight to that location and find the value. This makes it great for something like looking up a number – a computer won't forget a phone number in a hurry. By comparison, your brain does not hold a memory in a single location, nor does it have a direct way of going to a particular value. The way information is held is structured as patterns and images, which is why your brain may have trouble with a phone number, but it finds it a lot easier to recognise a face than a computer does.

Remembering how it's done

There are also several distinct kinds of memory in the brain. The lowest level is procedural memory, the memory that tells you how to do something. This takes place in the most primitive part of the brain, the part most closely shared with the widest range of animals, specifically the cerebellum and the corpus collosum, the bundle of nerves that links the two halves of the brain.

Procedural memory is accessed significantly more quickly than higher levels of memory, and with no conscious effort. If you are a touch typist like me, it's easy to demonstrate that procedural memory is different from conscious memory. As I type this, I am not looking at the keyboard and I don't think about where each key is. I simply think the words and my fingers type it. Procedural memory handles where to put my fingers and when to press.

If I try to remember where a particular letter, an N, say, is on the keyboard, I can't. I couldn't tell you. But I can type an N without thinking about it – my procedural memory knows the keyboard, but my higher memory doesn't. Something similar happens with experienced drivers. When you learn to drive you have to consciously be aware of what to do; how and when to change gear and so on. With experience, this ability becomes tucked away in procedural memory and happens without you having to think about it.

Remembering stuff

The higher level of memory, the conscious level, is processed by a number of areas of the brain. It is broadly divided into short-term, or working, memory and long-term memory. The prefrontal cortex, behind the forehead, administers the short-term memory, while the hippocampus, a central area of the brain that is supposed to look like a seahorse (but doesn't!) manages long-term memories, though the memories themselves are distributed throughout the brain.

One of the big distinctions between short-term and long-term memory is that we control what is in our short-term memory – you can consciously keep something in those short-term slots – but we have no direct control over long-term memory. You can't just flag something for memory and it automatically stays – you have to work at it. This is unnerving, when you think about it. You presumably think of yourself as rational, and yet here is one of the most important functions of your brain, probably the aspect that most defines you as an individual, and you have no direct control over it.

The brain is a self-patterning system, a common natural phenomenon. The more you use a particular neural pathway in the brain, the easier it becomes to use that pathway. If you think of the connections between neurons as electrical wiring, the wiring gets thicker as it is used, which makes it easier to use it again. So constantly accessing a particular memory makes that memory easier to recall – the mechanism behind the importance of revision.

Under pressure, your brain depends more than usual on these well-trodden pathways, which is why when you want to be creative it is best to be relaxed and not under pressure to find an instant answer. This gives the brain the chance to make use of thinner, less-frequented connections, where new ideas can spring up.

I know the face

Because our memories don't work like a computer, it helps to manipulate information to make it more acceptable to our brains and more accessible in memory. If, for example, you want to remember someone's name, there's a very simple technique to fix it in your memory: take the name and make a visual image out of it. Make it as colourful, visual, graphic (and even funny) as you can. Then combine the image with a mental picture of that person.

Let me give an example. Twenty-five years ago, when I first came across this technique, I thought I would give it a try. I happened to go into a pharmacy that lunchtime and decided to remember the name of the first person I came across with a name badge on. She was called Ann Hibble, a name I remember to this day. The image I conjured up was a hippopotamus (a big, purple hippopotamus) rearing up out of the floor of the shop and nibbling the woman's toes. *An hippo nibbling* – Ann Hibble.

As was pointed out when looking at the left/right split of the brain, things like colour, movement and drama all engage distinctive functions in the brain. So using imagery with colour, movement and drama helps ensure that this aspect of the brain is made use of, as well as the other brain modules more naturally involved with words. Memories are stored across both sides of the brain, so every little helps.

This technique for remembering a name involves fooling the brain. You are pretending that you're doing something more like the tasks your brain originally evolved to do. Humans evolved to recognise patterns, images and pictures in the world around us, so by superimposing an image on the name we hide the words under the visuals and get our memory to accept them more readily.

I probably wouldn't still be able to remember the name Ann Hibble if I hadn't reinforced it regularly by retelling the story. One essential to getting something to stick in memory is rehearsing that memory – digging it out and revisiting it on a regular basis, thereby thickening the neural connections. The ideal is to repeat this process on a gradually lengthening scale; perhaps after an hour, a day, a week, a month, six months, a year … if you do this the chances are the memory will never leave you.

Take down my phone number

At least names can be associated with objects and images, but numbers are even more abstract, and even more alien to the brain. When first faced with a number, the initial problem is that your short-term memory only has a very limited number of slots. You can only think of around seven things at a time without something popping out and being lost. Unfortunately a typical phone number might have eleven digits, which is beyond the capacity of your short-term memory.

Here's a made-up phone number: 02073035629. Taken as eleven separate digits it is pretty well impossible to remember, which is why phone numbers have traditionally been broken up into chunks. If you can memorise a chunk of numbers as a single item, you can squeeze the whole thing into short-term memory, en route to memorising it fully.

I remember that tail from somewhere

Memory is, of course, not unique to human beings. Anyone with regular exposure to animals will be aware how much memory features in their behaviour. Even the humble goldfish is perfectly capable of remembering things. This is a shame in a way, because the myth that a goldfish has a three-second memory makes for excellent jokes: 'Just because I have a three-second memory, they think I won't get bored with fish food ... Oh, wow! Fish food!' Okay, not always excellent.

However, anyone who has kept goldfish will be aware that they can remember things perfectly well – for example coming to a particular part of the pond or tank in response to a prompt before feeding, and a TV show has managed to get goldfish to learn the route around mazes. The idea that they have a three-second memory is nothing more than urban myth, probably equating intelligence and memory in some way, where in practice there is very little link between the two.

The brain scribble

The human brain is, without doubt, our crowning glory, and one of the most remarkable ways that we extend the functions of our brain is through the use of writing. The amazing thing about writing is that it is a means for one brain to communicate with another – in the case of this book, my brain communicating with yours – where time and space are removed as barriers.

Natural communication is limited in these respects. Mostly animals and plants communicate in the here and now. With a few exceptions in chemical-based communications that linger, a message is produced, consumed and gone, never to return. But writing takes away this limitation. You can take a book off the shelf and read words that were written thousands of miles away or even thousands of years ago. It is quite possible for you to have more communication on your bookshelves from dead people than living, and the chances are that few, if any, of the authors live on your street. When you read these words it will be months or years after the moment (13.32 GMT on Tuesday 4 October 2011) when they were written.

Of course we now have many other ways to communicate that are more instant than writing, but often these messages don't overcome time the way writing can. Because they are written down, these words will still be here in ten years time, maybe even a hundred years or a thousand. The cold call I just received from a stockbroker in New York was instantly consigned to the bin of time – the communication is as dead as a wolf's howl (thankfully, in this particular instance).

Writing has been crucial to the development of our technological society. Without writing there would be no science, only myth. With no way to build practically on the experience of previous years, we would always be re-inventing the wheel. Computer technology is often seen as something of an enemy of writing – why read a book when you can watch videos on YouTube? – but without writing there could be no computer software, no development of the hardware, and much of the content of the internet remains word-based.

Writing with pictures

Writing in the broadest sense is an extension of our brains. It is a way of taking information one human brain and storing it so that it can be revisited by another brain elsewhere in time and space. Originally, this was in the form of pictures. The cave paintings showing human beings, animals and patterns of hands dating back 30,000 years or more are not abstract daubing, but a means of communicating. They were fixed in space, slow to produce and difficult to interpret, but no one can doubt their ability to survive through time.

Over many years, straightforward pictures developed into pictograms. These still featured recognisable images, but the pictures were more stylised, making them quicker to execute, and more consistent in appearance. One pictogram would typically represent an object or, more subtly, a concept. It doesn't take a genius to decode a pictogram message showing fruit lying on the ground, then a pair of arms, then fruit in a basket.

The problem with a system like this is that there are too many symbols to cope with. A simplification would be to have separate symbols for fruit, basket and ground, and by drawing them in a particular relationship – perhaps with a special linking mark to suggest 'on' or 'in' – to combine those symbols. Now those simple pictograms are evolving into 'ideograms' – symbols that can put across an abstract concept like 'on'.

This is the stage at which 'proto-writing', the immediate ancestor of true writing, is thought to have emerged. Somewhere between nine and six millennia ago, symbols were being used with a degree of visual structure to put across a simple message. It is difficult to say when this proto-writing emerged, but many archaeologists believe the best early example currently known is that on the Tărtăria tablets, found in the village of that name in central Romania (once part of Transylvania). The clay tablets, a few inches across, feature just such a combination of stylised drawings, symbols and lines. It's possible that these were purely decorative, but everything about them suggests a message, a clear attempt to communicate information from one human brain to another.

Did you hear about my mummy?

Egyptian hieroglyphs form the best-known writing system that takes the next step – still using pictographs and ideographs, but in a much more formalised setting. The big advance here is that the pictures sometimes represent words, sometimes parts of words. Although hieroglyphs are the instantly recognisable script of ancient Egypt, they were only for special purposes. They were slow to produce and not well suited to, for instance, keeping accounts. A second system, hieratic, developed alongside hieroglyphs. It was also based on visual symbols, but was significantly more like a modern written script.

The Egyptians weren't the first to use true writing. Another of the regional superpowers of the time, Sumer, had what was probably the first written language, a cuneiform script – one where the characters are built up from wedge-shaped marks – a bit like the side-on view of a tack – made with the end of a stylus. To look at, these little symbols seem little more than a tally. But they are far more. They are a means of expanding the human brain, spreading information from one person to another.

By around 4,000 years ago, writing was spreading like wildfire. The Chinese system dates back to this time, using a large number of symbols (around 5,000) that represent words or parts of words. Our own alphabet has a ragged history before reaching our current written form. The name 'alphabet' shows its Greek origins (alpha and beta being the first two characters of the Greek alphabet), though the characters we use have a more complex history.

Abjads to alphabets

The earliest known predecessor of our script is the proto-Canaanite alphabet. Technically it was an abjad, an alphabet without vowels. The vowels are either implied by position, or marked using small change marks, like accents. This written form was used in the Middle East from around 3,500 years ago and was taken up by the Phoenicians. The symbols were adapted for both Greek and Aramaic lettering. Greek is thought to be the first true alphabet with vowels treated similarly to consonants, developing around 3,000 years ago.

Our own Latin or Roman script was derived over time from the Greek, and just as the US and the internet have spread the use of English today, the Roman Empire spread the use of Latin lettering as their language became a common tongue, one that that would outlast the empire itself by over 1,000 years. Isaac Newton's greatest work, *Philosophiae Naturalis Principia Mathematica* was written in Latin as late as 1687, while his *Opticks*, first published in 1704, though written in English, was translated into Latin for a wider audience.

It sounds capital

The Roman script we are familiar with – our capital letters with a few omissions – were the Roman equivalent of hieroglyphs. They were primarily used for carving in stone and important proclamations. Everyday writing used a different script called Roman cursive that looks part way between the capitals and modern lower case. Initially the letters varied greatly in size and placement, but over time they became more standardised in scale and more like our current lower case letters. Originally, though, capitals and cursive were two distinct schemes, and a writer would use one or the other, but gradually capitals crept in for emphasis in the midst of the cursive.

Exactly how capitals were to be used took a lot of time to settle down. In English, for instance, there was period when they were only used to emphasise new sections like sentences, then a time when, like modern German, they were used on pretty well every noun, before settling on the current compromise. It wasn't until printing came along that the two types of lettering would be called upper and lower case, referring to the moveable type that was used in printing until computerised printers became the norm. A page of print was bound together from a collection of individual letters on metal blocks. Capital letters would be kept in higher drawers or cases, while the 'minuscule' letters were stored in the lower cases.

Now we can see the true power of writing in enabling human beings to benefit from the power of their unique brains, taking them far beyond the capabilities of any natural creature on the planet. Think what you can do, thanks to writing: you can consult the wisdom of someone long dead, expanding the capabilities of your brain. You can support the working of your body by going online and purchasing something to eat from the other side of the world – or jot down a reminder on a Post-it note to ensure that you remember to do something important. And that's just your direct use.

Hardly anything around you that makes you different from your ancestors 100,000 years ago would exist were it not for the written word being present to aid its development. The written word makes it practical to have law, science, and literature, to name but three. Of course before the existence of writing there was an oral tradition – there were storytellers, for instance, but the difference in capability brought about by writing and its impact on human beings was immense. Speech can do a vast amount, but when a topic gets too complicated, writing is necessary to back it up.

The written word is immensely powerful, and many of us feel that it has a kind of magic. There is something special about books and bookshops, something very physically satisfying about handling a book. (Equally there's something special about a web search engine like Google, but that's a different kind of magic.) Of course, as a writer I would say that books are special, because books are what I do, but still, it's not an uncommon feeling. When written words are combined with a human's practical ability to make things happen they are of almost limitless power.

Are you human?

Impressed though you should be with the power of your brain that leaks out into writing, there are some aspects of your mental capability that can be emulated by a computer. As already mentioned, any old PC is much better at arithmetic and probability than any of us will ever be, and computers are also able to beat chess grand masters. In other cases we are just about holding our own – take, for example, the Turing test.

This was a trial devised by code-breaking and computing pioneer Alan Turing as a way of telling whether computers had finally come of age and were rivalling humans for intelligence. If you could sit in one room and communicate down a wire with 'someone' and couldn't tell whether that 'someone' was a computer or a person, then you could consider that the computer had achieved a form of artificial intelligence.

Over the years various programs have been written to try to interact convincingly with human beings, with various levels of success.

Experiment – Talking to computers

Go to **www.universeinsideyou.com**, click on *Experiments* and select the *Talking to computers* experiment. First try out Eliza, built into the page. This is one of the oldest computer programs designed to hold a conversation, written in the mid-1960s. It acts like a psychotherapist, echoing your statements back to you. It's quite easy to mess up, but if you play the game and don't try to be too clever, it is surprisingly good.

Then scroll down and click the link to take a look at Cleverbot. This is one of the best modern 'chatbots',

---

> as such programs are called. Even Cleverbot is rela-
> tively easy to confuse, but it has many more tricks up
> its sleeve than Eliza to attempt to look human.

At the Techniche Festival in Guwahati, India in 2011, the Cleverbot chatbot beat the Turing test. Or at least that's what has been claimed. In the test, 30 volunteers typed conversations, half with a human, half with a chatbot. Then an audience of 1,334 people, including the volunteers, voted on which conversations were with humans. A total of 59 per cent thought Cleverbot was human, making the organisers (and the magazine *New Scientist*) claim that the software had passed the Turing test.

By comparison 63 per cent of the voters thought that the human participants were human. This process can be a bit embarrassing for human participants who are thought to be a computer. I don't think, though, that the outcome is really a success under the Turing test. The participants were only allowed a four-minute chat, which gives the chatbot designers an opportunity to use short-term tactics that wouldn't work in a real extended conversation, the kind of interaction I envisage Turing had in mind.

And then there's the location of the event – a key piece of information that is missing from the published report is how many of the voters had English as a first language. If, as I suspect, many of those voting did not, or spoke English with distinctly non-Western idioms, their ability to spot which participants were human and which weren't would inevitably be compromised.

Would you kill to save lives?

Holding a conversation is one thing, but dealing with ethics is another. It's hard to imagine programming a computer to have a true understanding of ethics. After all, we don't necessarily even have a clear view of our own ethics. The theory may be straightforward, but when it comes to practice, it's easy to make decisions that it is then difficult to justify. Here's a well-known example ...

Imagine you are in a railway control centre and you see that a runaway train is headed down the track. It is totally out of control – you can do nothing to stop it. It is heading for a set of points where you can choose which of two lines it will travel down. If you do nothing, it will head down line A and plough into and kill twenty people who are on the track celebrating the opening of a new railway charity. If you throw the switch it will head down line B and will kill a single individual who is clearing the line of rubbish.

Let's be clear here: if you throw that switch you will directly cause someone to die who otherwise would not, but if you don't throw the switch, twenty people will die. What do you do? Decide before reading on.

Now let's slightly change the situation. Now you are standing on a bridge over the track. As before, a runaway train is heading down the track towards twenty innocent people, who will be killed if it isn't diverted at the points. You can't get to the controls, but there is a pressure switch beside the track below you that will flip the train onto a safe line, where no one will be killed. The only way to activate that switch is to drop something weighing twice your weight onto it. Sitting precariously on the parapet of the bridge is a very large person ...

If you push that person off the bridge, where they will definitely be killed by the train, you will save the other twenty people. If you do nothing, the twenty people will die. What do you do?

The majority of people would press the button to divert the train and kill one person rather than twenty. But many could not bring themselves to push the person off the bridge, even though this is apparently making exactly the same sacrifice.

Psychologists will tell you that this is because you are ethically capable of killing someone remotely by throwing a switch to save others, but, apparently entirely illogically, you can't face making the hands-on gesture. They point out that the same sort of shift has happened in warfare as the human technology used to kill each other has gone from hand-to-hand combat to bullets and missiles. However, I personally think that, useful though this thought experiment is for getting insights into our ethical systems, it is flawed.

The trouble is that the two different railway scenarios are not equally plausible. The first example genuinely could happen. It would be entirely possible to throw a switch and transfer a train to a different track, killing one person instead of twenty. But it seems very unlikely that you would have a pressure switch that required twice your weight and that you happened to have a person sitting there and knew what they weighed. The original phrasing of the problem, the form which has been actively used in psychological testing, is even more implausible.

It suggests that the person on the bridge is so fat that they can stop the train with their weight alone, showing a very poor understanding of physics on the part of the psychologists.

Worse than that, though, the psychologists are forgetting the impact of probability. The first test is not just more plausible as a scenario, but you could be happy that, technical failure excepting, when you throw the switch in the control room, the train will be diverted down the second track. However, even if you have been told that it will work, there is every possibility that pushing someone off a bridge could go wrong; they might fall in the wrong place, for example. The high level of uncertainty in the second test means that it would be much less appealing, even if there were no ethical concerns about taking a direct action to kill someone.

Trusting and ultimatums

Another experiment that you can carry out yourself gives powerful insights into the way we trust and also how we balance logic and emotion in our decision making – something else computers have problems with. We make decisions all the time, and this game really gets to the heart of what's happening when we make a choice – because it's not as simple a process as it may seem. The experiment is called the ultimatum game.

Experiment – Ultimatum game

Try this out next time you have some friends around to experiment on (or when you are next down the

---

pub). You need two people and a small amount of money, which you have to be prepared to part with for the sake of the experiment.

Explain to the two people you want to carry out a simple experiment. You are going to ask each of them to make a decision about some money. They must not discuss their decision in any way. Put the money on the table in front of them, so it is clear and real. Explain that you are going to give them this money to share – there are no strings attached, simply a decision to make.

The first person has to decide how the money is split between them. He or she can split it however they like. The money can be split 50:50, the decider can keep it all, or they can split it any other way they like. (It helps if you make the money a nice easy amount to split this way.) The person deciding must not talk about the decision in any way, but merely announce how the money will be split. The second person will then say either 'Yes' and the two of them will get the money, split between them that way, or 'No' in which case neither of them gets any money.

This game has been undertaken many times in many circumstances. The logical thing for the second person to do is say 'Yes', as long as the first person gives them something. Even if they're only offered a penny, it's money for nothing. In practice, though, the second person tends to say 'No' unless they get what they regard as a fair proportion of the money.

What counts as a fair proportion varies from culture to culture. Some will accept as low as fifteen per cent, others expect a full 50 per cent. In Europe and the US we tend to expect around 30 per cent or more before saying 'Yes'.

What the experiment shows is that we consider trust and fairness worth paying for. We are willing to lose money in exchange for putting things right. If human logic were based purely on economics then this just wouldn't make sense – you should always take the money. But your brain makes decisions based on a much more complex mix of factors than finance alone.

This is not to say that finance doesn't have some input into the complex system of weightings that is involved in decision making. If, for example, a billionaire decided to play this game, and offered a total stake of ten million pounds, the chances are you would happily accept being offered just five per cent – £500,000. Unless you are also extremely rich, half a million pounds is just too life changing an amount to turn down in order to teach someone a lesson and punish their lack of fairness.

It's an interesting exercise to think to yourself just how little you would accept in such circumstances. Where between £500,000 and £1 (which most people would reject) would you draw the line?

Weighing up the options

This kind of game seems to directly reflect the way the brain's decision-making capability functions. Different components of the decision are given weightings. The bigger the weighting the more important the factor is to the decision. These weighted values are then added together and whichever option gets the biggest weight wins. In the case of the ultimatum game, factors that are likely to be given weights include:

- How much money is involved?
- How much money do you already have and feel you need? (So, how important is the sum offered to you and your life?)
- How fair is the split that the other person comes up with?
- Is it for real? (Will you really get the money, or is it just hypothetical?)
- What is your relationship to the other person?

If a computer were undertaking this it would literally be multiplying the scores by the weightings, producing a set of numbers to compare. In the brain this kind of scoring is undertaken in an analogue fashion – it's more about the strength of an electrical impulse or the concentration of a chemical – but the effect is very much the same.

Allowing for all the factors

We like to think that we make logical decisions. Not the kind of cold logic deployed by Mr Spock, which would go for the money every time in the ultimatum game, but a more human logic that considers relationships, trust and fairness to be important as well as finances. And provided we do take into account all the factors that are coming into play, many human beings probably are logical in this fashion. But it's easy to miss what's really influencing a decision. The result can be an outcome that doesn't make any sense in terms of, say, your long-term good, because your decision-making process gave greater weight to short-term pleasure.

You can see this happening all the time from relatively mild personal decisions – like whether or not to have that tasty but unhealthy junk food or chocolate bar – to truly life-threatening decisions involving taking hard drugs or undertaking a high-risk activity. Human beings are not very good at factoring long-term impacts into our decision making. We can be aware of these factors, we can know very well what the implications are, but short-term gain very often outweighs long-term benefits.

Economists have traditionally been particularly bad at understanding human decision making. They used to expect perfect rational behaviour, where 'perfect' and 'rational' are defined as being behaviour that opti- mises the financial benefit to the individual. But such an approach is naïve in the extreme when thinking about real human beings, as is now increasingly being realised.

It could be you

Just take the simple example of playing the lottery. You are extremely unlikely to win a major lottery. The chances are millions to one against (to be precise, 13,983,816 to one in the UK Lotto draw) – similar to your chance of being killed in a plane crash, or being struck by lightning. Yet every week lots of people take part. What's going on?

In part it reflects our inability to handle probability. Just imagine if one day they drew the lottery results and the balls came out 1, 2, 3, 4, 5, 6. There would be an outcry. At best it would be assumed that the drawing mechanism was faulty and at worst it would be thought that there was fraud. There would probably be questions asked in parliament. And yet that sequence of numbers has exactly the same chances of being drawn as the numbers that popped out of the machine last Saturday. (In practice they were 29, 9, 15, 39, 17, 30.)

It's only when we see a sequence like 1, 2, 3, 4, 5, 6 coming out that we realise just how unlikely the chances are of winning – those astronomical odds don't really make a lot of sense to our mathematically challenged minds. However, despite this poor ability we have to cope with such numbers, the mathematicians, scientists and economists who regularly call people stupid for playing the lottery entirely miss the point as well. They are using a very poor model of human decision making.

I think I understand probability reasonably well, for example, but I still play the lottery. Admittedly in a controlled way with a small set monthly budget, but I do play. So why do I do it? It involves the kind of rewards that conventional economics is not good at reflecting.

If the sum involved in playing is so small that I can consider it negligible (perhaps the equivalent of buying a weekly drink at a coffee shop), then I can easily offset the almost inevitable loss against a very low chance of winning an exciting amount. To add to the benefit side of the equation, with this style of play I get a small win roughly every couple of months. This will inevitably be for between £3 and £10, but there are still a few minutes of delicious anticipation after getting the 'Check your account' email from the National Lottery when it could be so much better.

One of the important factors in considering the decision to play to be rational is that I totally forget about my entry unless I do get one of those emails. I don't anxiously check my numbers. I don't know what my numbers are. As far as I am concerned, once the payment has been made the money has gone, just as if I had spent it on those coffees. That way, any win is pure pleasure, because it has no cost attached to it. Let's face it, the only thing I'm likely to get the day after a visit to Starbucks is indigestion. (This is not casting aspersions on Starbucks. It's just that although I like real coffee, it upsets my stomach.)

Economics gets it wrong

Decision making based on finance alone ignores any enjoyment gained. In fact it ignores any benefits other than hard cash. If you took such an approach in your normal life, you would never spend any money on anything that hadn't got a clear financial benefit. Okay, you would buy food because you need to stay alive, but obviously you would select the cheapest food to give the necessary nutritional value. You would never go to the cinema, or theatre or to a concert. You would never buy a present or a treat. You would never eat in a restaurant, because you can always make something cheaper at home. The economist's 'perfect' life isn't worth living.

Did you do that consciously?

We have seen that you make decisions all the time to do things based on this complex mix of benefits, often with a skewed view towards short-term gain. But on the whole you probably think that your decision making is conscious. It's the 'you' inside your head, your conscious mind, that you assume is making the decisions.

When you are thinking about something – this question for example – where does that thought seem to take place? Where do you imagine 'you' to be located?

If you are like most people you will locate your conscious mind roughly behind your eyes, as if there were a little person sitting there, steering the much larger automaton that is your body. You know there isn't really a tiny figure in there, pulling the levers, but your consciousness seems to have a kind of independent existence, telling the rest of your body what to do.

This simple picture of your conscious mind as something inside your head, that pulls (imaginary) levers to make your body act faces one big problem. Modern brain studies show that a frightening amount of your actions are actually controlled by the unconscious mind. It's still 'you' that makes the decision – but not the conscious you, the active bit you think of as being in charge.

Let's imagine you are sitting outside with a ball alongside you. You pick up the ball and throw it. What happened in your brain? The natural assumption is that your conscious mind thought 'Okay, I'm going to throw this ball,' signals were sent through your nervous system, and your arm did the job. I'm not suggesting you literally had to consciously, if silently, verbalise 'Okay, I'm going to throw this ball,' but you made the conscious decision to do it, then it happened.

Brain activity is associated with increased blood flows. By monitoring brain activity using fMRI (functional magnetic resonance imaging) scans detecting blood flows in the brain, it's possible to see when the decision to perform the action is taken. This typically happens in the unconscious mind about a second before the hand begins to move. The conscious awareness of the decision takes place about one third of a second later. So, before you think 'I'm going to throw this ball,' your brain knows it is going to do it and is getting fired up. Only then do you become aware of the decision.

This sounds weird and rather scary. The decision is made before you are conscious of it. It's almost as if you were a kind of robot, with no true free will. But in reality it's more complex than this. Firstly, there is time for your conscious mind to abort the action. In the unlikely event you find yourself starting to do something you don't really want to do, you can stop it. And more significantly, it's not some alien external force that makes the initial decision, it is still you. You just aren't conscious of it.

Even so, this unconscious decision making does emphasise how complex our brain activity is, and how it really is very difficult to be definitive about how consciously individuals decide to do things (and perhaps to what extent they should be punished for doing badly or rewarded for doing well).

Mood swings and comfort breaks

A significant difference between the brain and a computer is that the brain is influenced much more by the environment it works in. You might think that your computer occasionally gets into a bad mood, but in reality, software glitches apart, it will make the same decision every time if presented with the same data. Your brain is much more likely to change its assessment due to outside influences.

One obvious example is mood. It's all too easy to make a bad decision simply because you are in a foul mood and are prepared, as the confusing saying goes, to cut off your nose to spite your face. You will make a decision that is bad for you, simply to irritate someone else or to be difficult. Surprisingly, two pieces of research in 2011 also identified that the state of your bladder has an influence on your decision making.

One paper, describing the rather unfortunately named concept (bearing in mind we're dealing with bladder control) of 'inhibitory spillover,' explains how, when we're under pressure to urinate, we do better at decisions where self-control is important. It's as if the fact that you are exerting conscious control over your muscles means that you also have better control of what would otherwise be knee-jerk decisions. These could be anything from making a high-speed identification of someone to taking financial decisions that result in short-term benefit but long-term problems.

The other piece of work shows that a full bladder isn't always a good thing: it can also make for bad decision making. As anyone who has tried to drive while desperate for the loo can confirm, this study found that we find it harder to pay attention and to keep information in our short-term memory when faced with an overfull bladder. This means an increased risk of having an accident when under such pressure.

It might seem that these two pieces of work are contradictory, but your brain is complex enough for these two results to be complementary in outcome. The fact that you find it harder to concentrate and retain information in memory when you have a full bladder is likely to reinforce a tendency not to jump in and make impulsive decisions, but rather to take a step back and exert more self control. This is fine when you have plenty of time, but not when making constant, important decisions. It's probably best, for example, that airline pilots and truck drivers have regular comfort breaks.

The brain's own painkillers

We also need to consider just how important the brain is when it comes to feeling pain. Although we associate pain with the area where we are hurt, the feeling of pain is generated by the brain, which means the brain can also turn off this feeling. We've seen earlier the way that swearing (page 43) and aspirin (page 146) can relieve pain, but another surprisingly effective approach is the use of placebos. These are dummy medicines with no content, usually sugar pills, which are used to test the effectiveness of new drugs. If a drug does no better than a placebo, it isn't worth using.

However, it has been known for a long time that placebos do themselves have positive effects. If your brain believes that the pill you are taking will have a beneficial effect it often will. This is particularly true of pain relief. The brain has its own natural ways of switching off the pain signals, and these can be encouraged into action with a placebo. In the case of pain relief, what a placebo does is make the brain assume that pain levels will reduce, and the brain makes this a self-fulfilling prophesy by releasing natural painkillers like the morphine-related endorphins.

Homeopathic misdirection

This appears to be the way that many alternative medicines work. Homeopathic treatments, for example, make no sense as an actual medicine. Homeopathy is supposed to work by combining an outdated medical idea that taking a small dose of a poison causes benefit and a magical concept that because something is similar to something else, it will have the same effect. So you take a small dose of a poison that causes similar symptoms to the one you are suffering and the result is to alleviate the suffering.

This makes no medical sense, and in practice homeopathic treatments are diluted so much that there will rarely be a single molecule of the original active substance in the liquid that is then dripped onto a sugar pill. The result is that a homeopathic pill is exactly the same as a placebo, and similarly it can have good effects by encouraging the brain of the person taking it to make things better.

Some supporters of homeopathy argue that it can't be a placebo as it also works with some animal problems, and the animals can hardly be fooling themselves as they have no idea what is going on. There seem to be three factors here. A proportion of the animals would get better whatever was done, but the owner would assume the remedy helped. Other owners fool themselves into thinking that an animal has become more comfortable (you can't actually tell the level of pain it is feeling, for instance), and finally an owner may well combine giving the treatment with extra care and attention, which itself will have a positive placebo effect on the animal.

The same applies to many other alternative treatments – acupuncture is a good example where there is little evidence of the treatment having any real benefit over and above being a placebo.

The ethics of placebos

The interesting thing here is whether or not this means that these treatments, or treatment using an explicit placebo, should be used. Many scientists have the knee-jerk reaction that they are unethical. To make effective use of a placebo (whether labelled alternative medicine or substituting for conventional medication) the person giving the treatment has to lie to the patient. It involves deception or self-deception.

The difficult ethical question is whether or not it is acceptable to deceive people in order to make them feel better. The placebo effect can be quite powerful, and is less likely to have side effects than many conventional medications. But is it possible to justify using deception to achieve positive results? Do the ends justify the means?

You might feel one answer would be that it ought to be justifiable, as long as it is cheap. After all, a lot of medicines are expensive. Given that a placebo (or a homeopathic treatment) is just a sugar pill, a bottle full should only cost a few pence. This might seem a way to justify what would otherwise be a cruel deception

Unfortunately, research has also shown that expensive placebos work better than cheap ones, when the people taking them are aware of the cost. When test subjects were given two placebo painkillers, one costing $2.50 per pill and the other $0.10 per pill, then given electric shocks, the subjects on the more expensive sugar pills experienced considerably better pain relief.

What is certainly true, though, is that the deception could be justified for the use of placebos and alternative medicine if it had a clear benefit and no disadvantage; there have, after all, been examples where such deception has resulted in suffering and death. Where a patient is given a homeopathic remedy or other alternative medicines to prevent malaria or 'cure' cancer, HIV and other life-threatening diseases, as has happened all too often, it can produce deadly results as well as raising false hope. If taking these remedies results in avoiding conventional treatment it can have terrible consequences, and deserves to be condemned.

A placebo is a mechanism that misleads the brain, using it to influence the body. Like all the ways your brain and body function, this mechanism has evolved through many generations. It's time to return to the mirror, to consider your body as a whole, and how it came to be here at all.

9. Mirror, mirror

Take a look in your mirror again. Try to forget that what you see is 'you', a human being. Just see an animal looking back from the mirror. An animal that isn't particularly different from an ape in appearance, even though your brain and the capabilities it gives you sets you apart. There was much talk in the early days of evolutionary theory of humans having descended (or ascended, depending on your point of view) from the apes – but that's a misleading picture.

Building your ancestor tower

To get a true picture of the evolution that produced your body in action we need to look back over your ancestry, all the way back to the earliest life on Earth that's related to you. Trying to imagine how you can get to a human being from something as simple as a bacterium, say, can be hard to visualise. Apart from basics like cell structure, containing water and DNA, it's difficult to see that you have a lot in common. But this is your heritage. A great way to picture how you got to that image in your mirror from these early life forms is to imagine building your ancestor tower.

We're going to represent you with a piece of Lego – specifically, a violet piece of Lego. You are on the top of a Lego tower. Below you is another violet piece of Lego. This is one of your parents (it doesn't matter which one). One of their parents is the next block down, and so on. Let's imagine we've constructed the whole tower, many kilometres high, that gives us a Lego block for each living thing all the way back to your earliest ancestor, the first life form in your ancestry.

How that first life form came into being is a different story, one we don't know the answer to. But let's stand well back and take a look at the tower you have built. It has a couple of cunning design features. The obvious one is the colour of the blocks. We have coloured them so that they show a rainbow. They run in colour from red down at the earliest days of your ancestry through to violet with you at the top. It's a complete rainbow.

How many colours in the rainbow?

When you look at a rainbow caused by raindrops when the Sun's out, it looks as if there are some distinct colours there. You can see a red block, an orange block and so on. But those divisions are totally arbitrary. The seven colours of the rainbow we talk about today were made up by Isaac Newton. Few people actually see seven colours in a rainbow, but Newton wanted there to be seven, probably to correspond to the seven notes in a musical scale. Even the divisions you can see are as a result of your brain fooling you, looking as usual for patterns.

In reality, the rainbow forms a continuity of colours, merging without any sudden change from red to orange, orange to yellow, yellow to green and so on. If you go down to the differences between the colours, based on wavelength of light, or energy of the photons involved, there are billions upon billions of colours. And that's what we have in the ancestry tower that culminates in your body. A *true* rainbow of colour.

No sudden changes

Pick any two adjacent blocks in the tower and to all intents and purposes they are the same colour, whichever blocks you choose. You will never see two adjacent blocks where, say, one is blue and the other is green. You will never see a transition from one colour to a different one. Yet over its height, the tower manages to shift in colour from red through to violet via all the other colours. There are, of course, very subtle differences between each brick, but they are far too small for your eyes to detect.

Similarly, with the creatures those blocks represent, each generation is, to all intents and purposes, exactly the same kind of animal as the previous one – you will never see a transition between one species and another. Each individual is the same species as its parents. Although your body is different from your same sex parent, the differences are largely cosmetic. You are the same species as your parents.

Look further back and there is no sudden break between human and prehuman, or, going further still, between a dinosaur/lizard-like creature and a mammal. Each time, the offspring is the same species as the parent, yet paradoxically we manage to get a shift from single-celled simplicity through to the ancestors of plants, fishes, dinosaurs, mammals and our fellow apes.

This is why the Victorian concept of a 'missing link' is so misleading. It suggests more of a change between generations than has ever happened. The term 'missing link' is, in fact, totally out of date. It refers to the idea that nature is composed of a great chain from the simplest forms of life (like bacteria) to the most complex (humans) and everything that has lived can be put into this structure – except that there are some missing links in the chain. The trouble with this picture is that it isn't possible to decide a sensible order for the chain. Is a hummingbird higher up the chain than a mouse? Is an earthworm higher or lower than a ragworm? It makes no sense.

A failure to link up

There was another subtlety in our tower design. Ordinary Lego has exactly the same pimples and holes on every block, so any block can clip into any other. On our ancestral blocks, the shape, size and number of pimples gradually changes with position in the tower. The differences between the block that represents your body and your parents will be indistinguishable. And we should be able to link a block with one many generations earlier. But if you run down the tower, eventually you will get to a block where your modern block will simply not link any more.

This is where the species boundary is *as far as you are concerned*. The ancestor block that yours couldn't click into, who we'll call Fred (but could be either sex), is a different species from you. You are incompatible. Biologically you would be unable to breed with Fred.

The really important, but rather puzzling thing about this is that we can't label Fred as the point at which a new species began. Hundreds of blocks either side of Fred are the same species as Fred; they can interbreed. It's just that Fred is a different species from *you*. What this demonstrates is that the whole idea of 'species' is a totally artificial one, devised by biologists before they understood evolution. The idea is useful as a marker, but it has to be seen as a relative one, not an absolute.

The babel of towers

The ancestor tower that produced your body doesn't stand alone. Every living thing has its own tower of ancestors. Some towers will be very similar to your own. A chimpanzee will have a tower that is identical until it gets to the point, near the top, that the human tower and the chimpanzee tower diverged.

That point where we had our last common ancestor with chimps really is closer than you might think. Your ancestor tower goes back over three billion years, but we and the chimpanzees split between seven and twenty million years ago. That means that only around 0.3 per cent of the Lego bricks in your tower differ from the chimp's. This doesn't mean human beings are in some sense descended from chimpanzees, or from any other existing ape. We both descended from a common ancestor that was neither chimp nor human.

To take another point of divergence, our common ancestor with mice lived around 75 million years ago. Go back down your ancestor tower to this point and you will find a small mammal that probably looks more like a mouse than a monkey, but it is neither. This time span, 75 million years, doesn't seem very long when you think that life has been around for around three billion years. It might not seem long enough to get from a mouse-like creature to you. But bear in mind the average generation time over those 75 million years might be around five years or less, which would mean at least fifteen million generations in which small changes could accumulate to make something very different.

There will be some ancestor towers that never make it to the present day – many of them, in fact. Think of the dinosaurs, for example. Each has just as rich an early part of the tower as you do, but stops short around 65 million years ago. (Note, by the way, that this was a similar time span to our divergence from the common ancestor we share with mice.) Other towers stop billions of years ago. None of these truncated towers represents a now-living creature.

Equally it is possible that there are ancestor towers that did not start from the same first block as our own. We don't know how life began on Earth, but if it happened once, it could have happened more than once, independently, in different locations. It seems likely, though, that every living thing yet discovered – animal or plant – does originate from the same first block. This is because every living thing we know so far has significant aspects in common. We are yet to discover a totally unique form of life that doesn't make use of a carbon-based structure and a DNA (and/or its related chemical RNA) mechanism as a control structure.

Proud to be 'just a theory'

The mechanism for moving up the tower is evolution. The body you see in the mirror is the product of a long evolutionary process. There is a lot of nonsense talked about evolution. It is sometimes attacked as being 'just a theory'. This is a fundamental misunderstanding of the nature of science – all of science is composed of 'just theories.'

If we take a fundamental bit of science, like Newton's laws of motion, we have a very simple set of rules that go something like this:

1. A body will remain at the same velocity (including stopped) unless acted on by a force.
2. The amount of force applied to a body is equal to the body's mass times the acceleration in the direction of the force.
3. Every action has an equal and opposite reaction.

Surely these rules aren't 'just a theory'? Well, yes, they are. The way science works is that a scientist or team puts together a hypothesis, which might be something like the laws above. They then test it against experiment: 'Do we find indeed this happening? Yes we do, so that strengthens the hypothesis.' The more evidence we have for it being true, the more likely it is that this is a useful theory. Once something goes from being just a hypothesis to being a theory, it has stood up well and is usable. But it could still be disproved somewhere down the line.

Newton gets it wrong

This has actually happened with Newton's second law from that list. Einstein's special relativity shows that if something is moving, the relationship between force and acceleration is more complex than Newton thought. And special relativity has never let us down yet; it's a better theory than Newton's. However, as it happens, the way Newton's theory fails doesn't make a noticeable difference most of the time. So we can still happily use Newton's simpler form in most circumstances.

Any theory can be disproved – it just requires some new evidence. And that includes theories that are given the misleading label 'laws'. No scientific theory can be absolutely proved, because a new piece of evidence could always come along that shows our assumptions to be wrong. But this doesn't mean science is no better than totally made-up ideas like magic. Science gives us the best picture, given our current information – it's just that it must always be a work in progress.

Evolution is a theory in the same sense that Newton's laws are. It could at some point be disproved, and our current best understanding is more complex than Darwin's original picture. Nonetheless it is at the moment by far our best theory, given the available evidence. In a way this is not surprising because it is such an obvious theory – in fact it's amazing that it wasn't discovered long before Darwin.

Evolving makes a lot of sense

The basis of evolution is very simple. You inherited various traits from your parents, making your body the way it is, as they did from their parents, and so on back down the ancestral tower. In Darwin's day they didn't know how this happened, but we now know it's down to genetics (and epigenetics). Some traits are likely to help a particular species to survive in its current environment. Others are likely to make it more difficult for the species to survive. Individuals that have the traits that help them

survive will be more likely to live long enough to reproduce. And so those traits are more likely to be passed on.

Over a long, long time, these gradual changes, coming
from a combination of the different mixes of DNA as different individuals breed and random changes occurring
in DNA as a result of mutation, will inevitably lead to
changes in the species. This is really all that evolution
is about – the way different generations are randomly
different from previous generations, combined with the
survival pressures of the environment.

Many of those who are unhappy with evolution and
would prefer that living things were designed by an
external force point out that this kind of change will
only result in gradual shifts within a species. It surely
couldn't result in, say, a fish-like creature evolving into a
human being. People who have this problem need to go
and play with an ancestor tower – as already mentioned,
there are no jumps from species to species. Each and
every generation is the same species as its predecessor.
That is the wonderful paradox of biology, brought about
by this arbitrary 'species' label. There don't need to be
any species-to-species leaps.

What use is half an eye?
Another problem those who are unhappy with evolution
have is that, bearing in mind that change happens very
slowly, what would be the advantage of a partly formed
change that didn't deliver any benefits? This is an issue
that plagued Darwin for some time. When you look at
your body in the mirror, it has many complex structures.
How, for example, could something as complex as an eye

come into being? How could you get from primitive creatures with no sight to something with a fully-formed eye?

This does not prove as much of a problem as it first appears. It might be that there is an intermediate stage that has a different benefit – for all we know, creatures with half-formed eyes might have looked more attractive to potential mates. But in fact, with the eye we know that there are straightforward benefits, because there are creatures out there right now with pretty well every intermediate stage between nothing at all and a complex eye. Some have light-sensitive patches on the skin; others have pinhole camera eyes – no lens, just a cavity with a retina; some have very crude optics; others have different variants on seeing, like an insect's compound eye, and so on.

Another example of a capability that seems to have little value if half-formed is having wings – either you can fly or you can't. But again, the reality is more subtle. With small wings, for instance, you might not be able to fly, but you can use them to get along a little more quickly when escaping a predator. And you may have some alternative use for them, like cooling yourself. It's entirely possible for a part-formed feature to have a different use that later gets discarded.

Part of the problem those who don't like evolution have with these complex structures is that, try though they might, creationists and others can't get the idea into their head that evolution is not being directed. So in asking 'Why would you have a partially formed wing?' there's an implicit assumption that evolution has a purpose, that it is working towards a wing. But evolution isn't like that – it is truly random, merely selecting along the way for things that are useful (or at least not a hindrance). Without the idea of some guiding principle being involved, there is less concern about these part-formed features.

Science can always be proved wrong

The problem with creationism and intelligent design, the alternative viewpoints usually put up against evolution to explain why your body is here – along with all the other facets of nature – is that they aren't science. Remember that the way science works is to test a theory against the evidence. But those who believe in an external designer say that there is no testable evidence for the existence of the designer – it is something that has to be taken on faith.

Most scientists will tell you that for a theory to be science, it has to be 'falsifiable'. That means that there has to be a mechanism for proving that the theory isn't true. One of the early scientific theories was that everything with weight tried to get to the centre of the universe, which was thought to be the centre of the Earth. It was wrong, but it was science. As more and more data became available from observing the Solar System and the universe around us it became clear that the Earth wasn't at the centre of everything. The theory was disproved. Similarly, evolution, quantum theory and relativity could all be disproved by the appropriate observations.

I'm not saying that scientists abandon their theories with good grace. Many cling on to them for a long time, until an overwhelming weight of evidence to the contrary forces them to admit their mistake. But a belief in a super-natural designer isn't like this – it can't be disproved. You can show that it isn't necessary, but you can't show it isn't true. I have to emphasise that just because it can't be disproved doesn't make it false, but it stops it from being part of the realm of science. Intelligent design and creationism aren't science and should not be taught as such.

Even some scientific theories suffer from this problem. Hundreds of scientists have dedicated their working life to string theory, a theory designed to explain the structure of all the different particles that make up the universe. But as yet no one has come up with a way of testing the theory (or any particular variant of it) and proving it false. Some argue that this means that as yet string theory is also not science. It is mathematics that may or may not have a link to the real world, but without that ability to test it and potentially disprove it, string theory must remain a second-class citizen as far as science goes.

The sense of wonder

With the paradoxical simplicity of evolution and its magical capability to change organisms from one species to another without ever seeing such a change from generation to generation, we have come to the end of our exploration of science using your body as our laboratory.

I hope that you will never again look in a mirror and just think 'I really need a bit more exercise.' Take a moment every time you see that remarkable structure to enjoy a sense of wonder. There's all of science coming together to make what you see work.

Your body is a window on the universe.

APPENDIX

Finding out more

Inevitably this book has only been able to skim through the areas of science that crop up in exploring the human body. Here are some recommendations if you would like to read more on a particular topic. The **www.universe insideyou.com** website lists all these books with links to read more about each of the books or to buy them.

A single hair
Human hairlessness
- *The Eternal Child*, Clive Bromhall (Ebury Press, 2004) – very effective theory of how the human ape become more like an infant to be able to survive in large groups, resulting in losing much of our body hair.

Atoms
- *Atom*, Piers Bizony (Icon Books, 2007) – a good mix of biography and science as we follow the trail to discover just what atoms are.
- *The Fly in the Cathedral*, Brian Cathcart (Viking, 2004) – brilliant story of the race to crack open and understand the atomic nucleus.

Temperature
- *Einstein's Refrigerator*, Gino Segre (Penguin, 2004) – (careful, there's another book of that name, and Segre's book is known as *A Matter of Degrees* in the

US), it's an excellent exploration of temperature, heat and cold.

Matter and energy

- *Why Does E=mc²*, Brian Cox & Jeff Forshaw (Da Capo, 2010) – getting from relativity to this famous equation isn't trivial. This book takes you through this and explains the standard model of particle physics. A little heavy in places, but very informative.

Antimatter

- *Antimatter*, Frank Close (Oxford University Press, 2009) – intriguing guide to antimatter: what it is, how it's made and how it's unlikely to be a serious component of a weapon.

String theory

- *The Trouble with Physics*, Lee Smolin (Da Capo, 2010) – absorbing exploration of the problems with string theory and how it has become an ineffective panacea that may not even be science.

Locked up in a cell

DNA

- *The Double Helix*, James D. Watson (Penguin, 1999) – a wonderful personal account of the discovery of the structure of DNA from one of those involved. Has been criticised for being very subjective, and was written in the 1950s, but still a great story.
- *Genome*, Matt Ridley (Fourth Estate, 2000) – excellent exploration of the human genome, each chapter

featuring a gene from one of the chromosomes. Very approachable.

Bacteria

- *Microcosm*, Carl Zimmer (Vintage, 2009) – fascinating study of the E. coli bacterium with plenty of lessons for the understanding of life as a whole, and our attitude to human genetic material.
- *Viruses vs. Superbugs*, Thomas Hausler (Palgrave Macmillan, 2006) – an intriguing but frightening look at one alternative to antibiotics: using phages, predatory viruses, to save us from killer bacteria that have become resistant to antibiotics.

Mitochondria

- *Power, Sex, Suicide*, Nick Lane (Oxford University Press, 2005) – it sounds like a political thriller, but it is in fact a fascinating exploration of the role of mitochondria.

Parasites, bacteria and other aliens in your body

- *The Wild Life of Our Bodies*, Rob Dunn (HarperCollins, 2011) – a contemplative exploration of the way predators, parasites and partners have shaped who we are today.

Neutrinos

- *Neutrino*, Frank Close (Oxford University Press, 2010) – small book on the hunt for these elusive particles that caused major headlines in 2011.

Through fresh eyes

Light

- *Light Years*, Brian Clegg (Macmillan, 2007) – the history of humanity's fascination with light from the earliest explanations to the latest theories.

The Big Bang

- *Big Bang*, Simon Singh (Fourth Estate, 2004) – although a little dated now on the alternatives to the Big Bang, still an excellent description of the origin of the theory and why it has so much support.
- *Before the Big Bang*, Brian Clegg (St Martin's Press, 2010) – the latest ideas on how the universe began, exploring the limitations of the Big Bang theory, looking at alternatives and if there can be a 'before'.

Astronomy

- *A Grand and Bold Thing*, Ann Finkbeiner (Free Press, 2010) – wonderfully told story of the effort to produce the Sloan Digital Sky Survey and how it has transformed astronomy.

Cosmology

- *The Fabric of the Cosmos*, Brian Greene (Allen Lane, 2004) – great exploration of the nature of space, time and matter, starting with relativity and quantum theory.
- *From Eternity to Here*, Sean Carroll (OneWorld, 2011) – the book *A Brief History of Time* should have been – really does explore the nature of time in the context of cosmology. Sometimes quite hard going, but brilliant.

- *The 4% Universe*, Richard Panek (OneWorld, 2011) – a useful and detailed history of the discovery of the existence of dark matter and dark energy, which make up around 96 per cent of the universe.
- *Afterglow of Creation*, Marcus Chown (Faber & Faber, 2010) – intriguing detective story, tracking back from the cosmic background radiation to the Big Bang.
- *Bang!*, Patrick Moore, Brian May & Chris Lintott (Carlton Books, 2006) – superbly illustrated basic introduction to cosmology. Probably works best for younger readers.

Quantum theory

- *Quantum Theory Cannot Hurt You*, Marcus Chown (Faber & Faber, 2007) – called *The Quantum Zoo* in the US, the best basic explanation of what quantum theory is all about.
- *The God Effect*, Brian Clegg (St Martin's Press, 2007) – the mind-boggling quantum entanglement explained, with plenty on the applications including unbreakable encryption, computers that can solve insoluble problems and matter transmitters.

Alien life

- *We Are Not Alone*, Dirk Schulze-Makuch & David Darling (OneWorld, 2010) – gives a real understanding of why we should be spending less on manned spaceflights and more on robotically exploring the possible life-bearing planets and moons in the Solar System.

Marching on the stomach
Chemistry

- *The Disappearing Spoon*, Sam Kean (Doubleday, 2011) – an entertaining romp through the chemical elements. Rather than take the kind of rigid, structured walk through the periodic table that might seem the natural approach, Kean lumps together a rather random collections of elements, linked only by the wonderful rambling tales of their discovery, use and general oddity.
- *The Periodic Table*, Eric Scerri (OUP, 2006) – without doubt the best book on the history and origins of the periodic table of the elements. It's hard work, not a light read, but if you really want to get a feel for where this amazing structuring of the elements comes from, this is the book to give it.

Aspirin

- *Aspirin*, Dairmuid Jeffreys (Bloomsbury Publishing, 2005) – the story of aspirin from quinine substitute to heart medicine. Excellent background and genuinely fascinating.

Feeling dizzy
Electricity

- *Electric Universe*, David Bodanis (Little, Brown, 2005) – excellent as a teen introduction to the wonders of electricity. Some adults may find it a bit gushing, but otherwise fine for older readers too.

Gravity

- *Gravity*, Brian Clegg (St Martin's Press, 2012) – an in-depth but approachable exploration of gravity, general relativity, quantum gravity, anti-gravity and more.

Time Travel

- *Build Your Own Time Machine* (*How to Build a Time Machine* in the US), Brian Clegg (St Martin's Press, 2011) – the real science of time travel explained.

Two by two

Genetics

- *The Selfish Gene*, Richard Dawkins (Oxford University Press, 2006) – although predating many of the discoveries in epigenetics, still an excellent introduction to evolutionary genetics.
- *Not a Chimp*, Jeromy Taylor (Oxford University Press, 2010) – convincing exploration of the very real differences between humans and chimps, overlooked by simply comparing the genes.

Epigenetics

The Epigenetics Revolution, Nessa Carey (Icon Books, 2011) – readable and insightful explanation of the way that genes and DNA are just the starting point, but to understand how humans (and other life) are formed, we need to go beyond the gene.

Dogs

- *If Dogs Could Talk*, Vilmos Csanyi (The History Press, 2006) – real eye-opener on the nature and sophistication of the mind of a very familiar creature: the dog.

Cloning
- *After Dolly*, Ian Wilmut & Roger Highfield (Little, Brown, 2006) – excellent combination of a history of Dolly the sheep with an exploration of cloning.

Mutants
- *Mutants*, Armand Leroi (HarperCollins, 2004) – truly remarkable book that uses human mutation to explain how we are all formed, while avoiding the voyeurism of the freak show.

Crowning glory
Probability and statistics
- *The Tiger That Isn't*, Michael Blastland & Andrew Dilnot (Profile Books 2007) – brilliant excursion into the way we misuse and misunderstand numbers and statistics, and how to see around our probability blindness.

The brain – why it gets things wrong
- *Brain Bugs*, Dean Buonomano (W. W. Norton, 2011) – entertaining exploration of the brain, finding out more about it from its failings.
- *The Invisible Gorilla*, Chabris & Simons (Broadway, 2011) – why perception lets us down, from the devisers of the basketball video, the asking directions experiment and more.

- *A Mind of Its Own*, Cordelia Fine (Icon Books, 2006) – short and very readable introduction to the many ways our brains deceive us, illustrated throughout by psychological experiments.

The mind and brain

- *How the Mind Works*, Steven Pinker (Penguin, 2003) – very approachable exploration of thought and the mechanisms behind it.
- *Incognito*, David Eagleman (Canongate, 2011) – hugely readable exploration of the way our brains handle sensory input and make decisions, showing how (relatively) little influence the conscious mind has.
- *The Brain Book*, Rita Carter (Dorling Kindersley, 2009) – surprisingly good adult picture book on the brain and how it works.

Memory

- *In Search of Memory*, Eric R. Kandel (W. W. Norton, 2007) – excellent account of the work of Nobel Prize winner Kandel, putting his studies of the cellular nature of memory in the context of his life.

Language and writing

- *Through the Language Glass*, Guy Deutscher (William Heinemann, 2010) – really engaging book on linguistics and what it can reveal about human perception. Don't be put off by the 'linguistics' word – not at all dry and dusty.

- *Why We Lie*, David Livingstone-Smith (St Martin's Press, 2004) – be amazed, not just at how much we lie, but how essential lying is for the operation of society.

Artificial intelligence and the Turing Test
- *The Most Human Human*, Brian Christian (Viking, 2011) – the author examines what makes us human as he becomes one of the test subjects in a human-versus-computer Turing test.

Placebos, alternative medicines and treatments
- *Trick or Treatment*, Simon Singh & Edzard Ernst (Fourth Estate, 2002) – superb analysis of alternative medicine showing how the early trials often quoted by supporters were often unscientific, and new data prove most to be no different from placebos.

Codes and ciphers
- *The Code Book*, Simon Singh (Fourth Estate, 2002) – the development of codes and ciphers through the ages with lots of historical context and interest.

Mirror, mirror
The origin of life
- *Genesis*, Robert M. Hazen (Joseph Henry Press, 2005) – very personal exploration of the possible origin of life from both experiment and field work.

Evolution
- *The Autobiography*, Charles Darwin (Icon Books, 2003) – not at all stuffy and Victorian: this short book

gives a fascinating insight into Darwin as a human being.

- *Why Evolution is True*, Jerry A. Coyne (Oxford University Press, 2010) – a persuasive and plain-spoken summary of the evidence for evolution by natural selection.
- *Here on Earth*, Tim Flannery (Allen Lane, 2011) – beautifully written introduction to evolution and the history of Earth and its inhabitants.
- *Written in Stone*, Brian Switek (Icon Books, 2011) – excellent exploration of how our understanding of fossils has developed over time and why science thinks the things it does about the development of animals on Earth.

Index

Please note: locators in italics refer to illustrations.